Philosophy and Medicine

Volume 134

The Philosophy and Medicine series is dedicated to publishing monographs and collections of essays that contribute importantly to scholarship in bioethics and the philosophy of medicine. The series addresses the full scope of issues in bioethics and philosophy of medicine, from euthanasia to justice and solidarity in health care, and from the concept of disease to the phenomenology of illness. The Philosophy and Medicine series places the scholarship of bioethics within studies of basic problems in the epistemology, ethics, and metaphysics of medicine. The series seeks to publish the best of philosophical work from around the world and from all philosophical traditions directed to health care and the biomedical sciences. Since its appearance in 1975, the series has created an intellectual and scholarly focal point that frames the field of the philosophy of medicine and bioethics. From its inception, the series has recognized the breadth of philosophical concerns made salient by the biomedical sciences and the health care professions. With over one hundred and twenty five volumes in print, no other series offers as substantial and significant a resource for philosophical scholarship regarding issues raised by medicine and the biomedical sciences.

More information about this series at http://www.springer.com/series/6414

Lisa M. Rasmussen
Editor

Human Guinea Pigs, by Kenneth Mellanby: A Reprint with Commentaries

Springer

Editor
Lisa M. Rasmussen
Department of Philosophy
University of North Carolina at Charlotte
Charlotte, NC, USA

ISSN 0376-7418 ISSN 2215-0080 (electronic)
Philosophy and Medicine
ISBN 978-3-030-37696-3 ISBN 978-3-030-37697-0 (eBook)
https://doi.org/10.1007/978-3-030-37697-0

This Springer imprint is published by the registered company Springer Nature Switzerland AG
The registered company address is: Gewerbestrasse 11, 6330 Cham, Switzerland

[Dedication from Kenneth Mellanby in the first and second editions:]

This book is dedicated to all the pacifist volunteers who have given their services as "guinea pigs" in medical experiments, and in particular to the volunteers who have worked with me at the Sorby Research Institute in Sheffield.

Foreword: Truth, Mites and Guinea Pigs

I'd like to begin by introducing a hypothesis. My colleagues and I have been studying the biology of bacteria that live in human armpits for several years. Armpits contain glands called apocrine glands arranged in a cluster called the "axillary organ." These glands are often mistakenly described as sweat glands. But they do not produce sweat. They instead produce a substance that serves no purpose other than to feed bacteria. The bacteria eat this substance and, in doing so, turn it into another compound which produces an odor or really a mix of odors (Gordon et al. 2013). It is these odors, these bacterial odors, that we smell when we note the underarm aromas of someone near us on an elevator or at our side during a seminar. Parts of what I've just told you have been known for many years. What was not known was how the use of antiperspirant affects these microbes, nor what consequences any such effect might have, nor why our bodies evolved glands that feed these microbes in the first place. My colleagues and I resolved one of these questions. We were able to show that the use of antiperspirant, which partially closes the apocrine glands, disfavors the slow-growing bacteria species that give armpits their characteristic odor. In their place grow a diversity of unusual, poorly studied bacteria and several common types of bacteria that are closely related to pathogens (Urban et al. 2016). The effect of antiperspirant on apocrine glands and the bacteria they feed appears to be part of what makes not only the armpit bacteria but also, more generally, the skin bacteria of western adults so different from those of other primates, including our close relatives, gorillas and chimpanzees (Council et al. 2016).

In other words, having been convinced by antiperspirant-producing companies to use antiperspirant, we are favoring many poorly known species, and we are doing so on a massive scale, in such a way that makes our skin unlike that of our living close relatives and likely unlike that of our recent ancestors. This is intriguing on its own, but it is especially intriguing if my hypothesis about the original function of apocrine glands is correct. I hypothesize that these glands evolved, in part, so as to feed specific, slow-growing bacteria that produce antimicrobial compounds that spread across our skin and aid in defense against pathogens. If my hypothesis is correct, when we use antiperspirants, we not only remove part of our defensive layer of beneficial microbes, we also inadvertently favor groups of microbes likely to themselves be pathogenic. I could be wrong. The beauty of a hypothesis is that it is an idea waiting to be disproven.

I've introduced the story of the armpit microbes and my hypothesis as a modern backdrop against which to consider the experiments Kenneth Mellanby conducted on *Sarcoptes* mites with conscientious objectors. The armpit microbe example serves as a reminder that every day in our modern lives we are part of a self-experiment of some sort, a self-experiment on our own health and well-being. Every day that we apply a new product or eat a new food, we are doing a self-experiment. We just typically do so in a way that doesn't generate any new data or truths other than those we anecdotally come to believe ourselves. Without the context of the scientific method, these anecdotes sometimes lead us toward useful insights. More often they lead us to ruin. The Internet is full of "insights" based on anecdotes – thousands of such insights, most of them absolutely wrong.

One way to wrest truth from anecdotes is to wait for people to embark, accidentally, on one or another kind of collective experiment and then to consider the results. This is what we would have if we looked at the health of people who do and don't use antiperspirants (which we could still do). But the trick with such studies is that they are typically correlative. What would we really know if we found out that people who used antiperspirants tend to be more likely to suffer from infections? We'd have a stronger inference, but it would be an inference, nonetheless. What we'd really want would be an experiment in which we could control who received a particular treatment and who did not. Again and again in the history of science, experiments have been required to convince scientists of a new idea. But should we do such an experiment? If we really imagine that antiperspirants cause people to be more likely to suffer infections, is it safe to ask people to use them for many years? And if we convinced people it was safe, could we ensure that the people who said they would use antiperspirants really did and those that said they didn't, *didn't*? Over months, sure. Over years or even decades?

It was a conundrum like the one we face with the skin microbes, a conundrum about how to do an experiment on humans and do it in a way in which he could ensure compliance, which led Kenneth Mellanby to enlist conscientious objectors as experimental subjects. The objectors volunteered, but did so in a situation in which they had relatively little power and money. In this way, they were a great deal like modern military recruits. Subsequent to the time of Mellanby's experiments, ethicists have helped science to decide, as a field, that experiments like those Mellanby carried out, in which relatively poor and powerless people volunteer their bodies on behalf of their society, are unethical. But we continue to consider military recruitment that targets and disproportionately attracts young people with the least power and puts them in harm's way to be, if not ethical, at least relatively unnoteworthy. Why we have decided differently in these two cases is beyond my scope here, but it is a contradiction Mellanby himself is likely to have raised.

The other thing Mellanby might bring up were he still alive, and indeed that he did bring up in the second edition of his book, is that the ethical challenges of a particular scientific study should be weighed against its benefits and against the challenge of arriving at truth in any other way. Many of the major insights made with regard to the management or effects of particular species on humans were made in the 1940s. Some of the most important advances in human nutrition were made during the same time period (Tucker 2007). There exist some questions that are nearly impossible to resolve without large-scale experiments on people. Interestingly, those questions seem to be becoming more apparent as we understand more about the complexities of the human body. Some of these questions are hard – but possible – to answer without experiments on humans. Others, though, are probably impossible to satisfactorily address without large-scale human experiments.

We can consider, for example, the case of microbes on the skin that I already raised. It would be hard to figure out just what the benefits and costs of the different skin microbes or the different products that affect those microbes are without experiments on humans. Yet, it is, I suspect, ultimately possible, just difficult. We can grow skin microbes in the lab. We can do some relatively small-scale studies on humans which we know have no negative consequences (not even discomfort). We can probably figure out most of what we need to figure out without a study in which we impose a particular skin treatment on some people and not on others for years. In part, this is because it seems as though most of the potential benefits of skin microbes are acute benefits. The microbes seem as though they might benefit us over the short time scales on which ethical noncoercive studies are possible. In addition, the microbes on human skin and the biology of human skin are sufficiently similar to those of nonhuman animals that some of the questions we'd like answers to we could study using mice, or, say, guinea pigs.

Things become more complex, however, if we consider the microbes not on the skin but instead in the gut. A series of recent studies – hundreds of studies – has begun to show that various factors influence the specific species of microbes any of us have in our guts and that the specific mix each of us has affects health and well-being. Your microbes have the potential to influence your risk of Alzheimer's, of dementia, of inflammatory bowel disease, and even of multiple sclerosis. Yet, even as we begin to understand these effects, it remains very difficult to say with any precision which specific mix of microbes is a healthy mix and which is not. The challenge here is much greater than with the skin, in part because the microbes in the gut are much more diverse (there are many more species that do many more different things), most of the microbes in the gut cannot yet be grown in the lab (we don't know what they eat), and the effects of particular microbes are contingent on the genetic background of the person they inhabit. In theory, we might be able to resolve all of these things to understand which microbes each of us need to live healthy, enjoyable, well-balanced lives. In practice, that may be impossible. In practice, there are so many potential species in our guts, and their influence might depend on so many different things that disentangling their relative influence may be something we are never able to do well. To really sort out the truth, we need a big experiment, in which peoples' microbes are cleaned out using antibiotics and then they are given different mixes of microbes that they live with for years, for decades. Such an experiment is obviously unethical and shouldn't be carried out. Yet, my guess is that without such an experiment, we will never understand just which microbes we need, when, and why, and as a result, we will spend the next decades buying products and probiotics to make our gut microbes healthier, the value of which is inferred from bits and pieces of science and personal anecdotes. Perhaps time will prove me wrong. I doubt it. If I'm right, we are forced to return to a question that Mellanby posed, in different ways, again and again: How many lives do we balance an experiment against when its consequences will be measured not just across one generation but instead against thousands? But there is a huge caveat here, even in posing this question. The caveat has to do with who decides which questions are the ones that can only be resolved with experiments on humans.

I've already noted that I think it is possible to understand the basic workings of skin microbes without an enormous, long-term human experiment. In part, this is because skin microbes can be grown in the lab and because the skin of humans is not so very different from that of other mammals (compared to the much larger differences in the guts of different mammals). But wait, if this is true of skin microbes, shouldn't it also be true of the mites that live *in the* skin, mites like those that Mellanby studied? Yes and no. Some of the mites that live on the skin, including the *Demodex* mites we study in my lab, are very specialized on their particular host. All adult humans appear to have at least one species of *Demodex* mites, and most people have two (Thoemmes et al. 2014). These mites live their entire lives on a human body and appear to, in most cases, have very little effect. However, these mites appear to be so well adapted to their hosts that they don't survive well in the lab (we've never been able to raise them) and the mites from one mammal species don't survive well on another mammal (Palopoli et al. 2015). In this way, these mites are like gut microbes: host-specific, finely tuned to the genetics and physiology of their hosts, and really challenging to study experimentally. Mellanby, however, did not study *Demodex* mites. He studied scabies mites which are very different from *Demodex* mites in that a single species of scabies mites appears able to colonize many tens of different species of mammals. There may be nuances that are currently being missed (some populations of scabies mites seem somewhat adapted to dogs, for instance), and yet human scabies mites, for example, are now thriving on wombats, to the great detriment of wombats (Arlian and Morgan 2017). As a result, the basic biology of human scabies mites, whether they live in bedding and how they transfer from one host to another, the basic biology that Mellanby studied, could almost certainly have been studied on another mammal. Our understanding of scabies mites was advanced greatly by Mellanby, and his work is still the work of record with regard to the treatment and transmission of scabies mites (see, e.g., Arlian and Morgan 2017). But it was research that could have been done without conscientious objectors. Mellanby didn't need human volunteers. He could have used dogs, or foxes, or even wombats. It would have been trickier in some ways, but it could have been done. He probably could have even used actual guinea pigs! Indeed, most of the work that has followed up on that of Mellanby, and aided in our understanding of how to control scabies mites, was not done on humans but on other mammals. This brings me to my final point, the challenge of trusting scientists.

I began this essay by sharing my hypothesis as to the function of apocrine glands in armpits and also the function of the microbes they feed on our skin. It is easy to imagine an entire field of study based on this hypothesis. I may, however, be entirely wrong. Some of the collaborators with whom I work on skin microbes think I probably am. Science relies on new hypotheses and on disagreements about whether or not they are true and also on disagreements about how best to study a particular question. Yet, when we make decisions as to the ethics of experimenting on humans, we need to do so in the context of as much certainty as possible. Scientists are often relatively unable to assess such uncertainty. They are also, very often, sufficiently thrown into the details of their own work that they don't know the history of the question they are studying as well as one might hope, nor the context of their question, much less the history and broader context in which questions about ethics should be decided.[1]

If we were to embark on a grand experimental study of human skin microbes and antiperspirant, or gut microbes and dementia, or some other such study, at some stage, we ultimately need to trust that the scientists are making the right decisions. But our understanding of the biological world is so modest (most species are not yet named; most named species have not yet been studied in any detail) that such confidence will typically be unfounded. The result is that we will always be balancing: balancing the ethics of a particular study against its potential benefits and human costs and balancing our confidence in scientists against the reality that scientists are wrong most of the time (even as the collective body of scientific research is ever more true). In the end, Mellanby's book is a fascinating look into the way science works, the challenges of wrestling truths from the universe, and the inevitably difficult decisions about which science we should and should not do. And also, of course, it is the story of a very unusual mite, a mite that is becoming more common again and that, despite decades of study, remains very poorly known.

Applied Ecology
North Carolina State University
North Carolina, USA

Rob Dunn

References

Arlian, Larry G., and Marjorie S. Morgan. 2017. A review of *Sarcoptes scabiei*: Past, present and future. *Parasites & Vectors* 10 (1): 297.

Council, Sarah E., Amy M. Savage, Julie M. Urban, Megan E. Ehlers, J.H. Pate Skene, Michael L. Platt, Robert R. Dunn, and Julie E. Horvath. 2016. Diversity and evolution of the primate skin microbiome. *Proceedings of the Royal Society B: Biological Sciences* 283 (1822): 20152586.

Currier, Russell W., Shelley F. Walton, and Bart J. Currie. 2011. Scabies in animals and humans: History, evolutionary perspectives, and modern clinical management. *Annals of the New York Academy of Sciences* 1230 (1): E50–E60.

[1] For example, although the authors in this volume appear to be in (reasonable) agreement that from our modern perspective that Mellanby's experiments were ethically problematic, a recent review of the biology of scabies mites heralded Mellanby's studies as an example of ethical human experimentation. See Currier et al. 2011.

Gordon, James A., Corrine J. Austin, Diana S. Cox, David Taylor, and Ralph Calvert. 2013. Microbiological and biochemical origins of human axillary odour. *FEMS Microbiology Ecology* 83 (3): 527–540.

Palopoli, Michael F., Daniel J. Fergus, Samuel Minot, Dorothy T. Pei, W. Brian Simison, Iria Fernandez-Silva, Megan S. Thoemmes, Robert R. Dunn, and Michelle Trautwein. 2015. Global divergence of the human follicle mite *Demodex folliculorum*: Persistent associations between host ancestry and mite lineages. *Proceedings of the National Academy of Sciences* 112 (52): 15958–15963.

Thoemmes, Megan S., Daniel J. Fergus, Julie Urban, Michelle Trautwein, and Robert R. Dunn. 2014. Ubiquity and diversity of human-associated *Demodex* mites. *PLoS One* 9 (8): e106265.

Tucker, Todd. 2007. *The Great Starvation Experiment: Ancel Keys and the men who starved for science*. University of Minnesota Press.

Urban, Julie, Daniel J. Fergus, Amy M. Savage, Megan Ehlers, Holly L. Menninger, Robert R. Dunn, and Julie E. Horvath. 2016. The effect of habitual and experimental antiperspirant and deodorant product use on the armpit microbiome. *PeerJ* 4: e1605.

Introduction: Kenneth Mellanby's Scabies Experiments[1]

Chasing footnotes during research on unregulated human subject experimentation, I came across a reference to *Human Guinea Pigs* – not the more familiar Pappworth volume by that name but an earlier book of the same name by entomologist Kenneth Mellanby.[2] Mellanby's title referred to experiments on conscientious objectors in England during World War II to study the transmission and treatment of scabies. The experimental findings had a profound effect on the treatment of the disease and a correlative effect on English troop strength during World War II. The significance of the dates of his research was obvious: this research had been conducted before the articulation of the Nuremberg Code or any other significant national standards of regulation or oversight of human subject research. For the purposes of philosophy and medicine, what is interesting about this work is Mellanby's sustained explanation of and reflection on how he treated human subjects of experimentation in the absence of these regulations and standards. An additional feature of interest is Mellanby's subsequent role as a reporter for the *British Medical Journal* during the Nuremberg trials, publishing controversial opinions about the value of Nazi medical experiments.

I found few copies available for purchase, most of them rather expensive. Given the historical significance of the volume (which Robert Baker discusses in his commentary in the volume) and the fact that we currently confront questions about unregulated research (e.g., citizen science, social media and corporate research, or biohacking), it made sense to find out whether we could bring the volume back into print in the Philosophy and Medicine series. This turned out to be a much more dif-

---

Kenneth Mellanby died in 1993.

---

[1] I would like to thank Jane and Alex Mellanby, Kenneth Mellanby's children, for granting permission to reprint this work. I am also indebted to my graduate assistant, Bailey Davis, for helping to edit and proofread the volume.

[2] The readers will see, in Mellanby's preface to the second edition, that Pappworth's title duplication irked Mellanby.

ficult task than I had anticipated and an object lesson in scholarship in the modern publishing era.

The first edition was published in London in 1945 by Victor Gollancz, and the second edition was printed by Merlin Press in 1973, probably motivated by the publication and reception of the Pappworth volume of the same name in 1967. However, Gollancz was later sold to Houghton Mifflin, which was later sold to Cassell, which was later acquired by Hachette. This history of sales made it challenging to track down the copyright ownership of the volume. It took months to get a reply from each contact I tried, and when they did respond, they could not confirm the current rights holder and suggested that I should ask someone else (often someone I had already asked). One suggested I try to find Mellanby's heir(s), but even based on the names given in Kenneth Mellanby's obituary, I was unable to find them.

So, I did what researchers sometimes do in the modern era: I googled Kenneth Mellanby's name and followed the breadcrumbs. Somewhere several pages into the search results, I found a genealogy page on the Mellanby name and a genealogy entry by someone who might be related. I emailed that address, and Kenneth Mellanby's niece responded. She in turn put me in touch with Mellanby's two children, who, I am grateful to say, graciously granted permission to reprint the volume. This process has made abundantly clear that modern copyright laws are not appropriate for the needs of historical scholarship.

II. Commentary Essays

Part 1 of the volume reprints the second edition of *Human Guinea Pigs* in its entirety. When it was published in 1973 as a second edition, the entire first edition was included, with very minimal edits for errors and nothing that affected the content. A second preface and Chapters 14–17 were added to the second edition. Part II of this volume presents several commentary essays highlighting for readers the scientific, historical, and ethical issues and contexts raised in *Human Guinea Pigs*.

In "Truth, Mites, and Guinea Pigs" (Preface), entomologist Rob Dunn considers the balance between the pursuit of truth on one hand and our ethical obligations to human beings on the other. As he points out, there are many experiments that we *could* do in the interest of science and for the long-term benefit of future generations, but the more important question is whether we *should* do them. He also sketches ways in which, even outside of the context of science, we are experimenting all the time (albeit in uncontrolled fashion) – for example, with the widespread use of a seemingly innocuous agent, antiperspirant, that appears to affect the populations of microbes on our skin, with unknown consequences. Mellanby's work was a cornerstone of what is known about the transmission of and treatment for scabies, but Dunn points out that Mellanby could have chosen to balance the ethics and science differently than he did.

In "*Human Guinea Pigs*: Mellanby, Pappworth and Club Regulation," Robert Baker gives a rich account of the differences between the two authors (Mellanby and Pappworth) who published under the same title. He shows how the insulation that comes with class and status can sometimes blind researchers to the ethical implications of how they choose to balance risks to humans against the pursuit of science. As Baker explains it, "Pappworth had the [outsider's] perspective to see what gentlemanly insiders were blind to: that British physicians were unethically exploiting their patients, using them as unconsenting 'human guinea pigs'" (Baker, this volume, 119). Mellanby's approach to the ethics of experimentation exemplifies something more akin to "club regulation," and as Baker notes, we should consider the tendencies of in-group or self-regulation as we think about criticisms of the current regulatory regime in human subject research.

Ana Iltis considers the force and impact of Mellanby's narrative – in contrast with his scientific papers – in "*Human Guinea Pigs*: Uncovering Principles for Ethical Research through a Personal Narrative." She highlights ways in which Mellanby's narrative evinces concern for some of the hallmarks of ethical human subject research, such as concern for scientific rigor, choice of topics of significant importance to society, and voluntary participation in research. By contrasting Mellanby's scientific papers with his book's first-person account, Iltis shows the ways in which important information is elided in a scientific "objective" publication, while the narrative style of *Human Guinea Pigs* allows for more nuanced ethical considerations.

In "The 'Untrammelled' Scientist and the 'Normal' Volunteer: Some Reflections," Alastair Campbell points to all that we don't know about Mellanby's experiments, by virtue of what he doesn't say in his book. With the benefit of his own family's experience with conscientious objectors (COs) in the UK during World War II, Campbell also highlights some of the ethical issues in using COs in research, focusing in particular on the issue of voluntariness. In addition, he describes an example of what he calls "lone ranger" research – in which the researcher is the ultimate judge of the appropriateness of the study – which, together with the example of Mellanby's research, shows us the potential dangers of human subject research without ethical oversight.

Joseph S. Brown and Toby Schonfeld, in "Compensatory Service for Conscientious Objection," consider conscientious objection in Mellanby's time, particularly with respect to the implication that objectors should serve in an alternative manner in exchange for being released from combat. They explore what implications this may have for conscientious objection in health care, in particular whether there are corresponding compensatory obligations in exchange for release from providing particular health-care measures. They suggest that compensatory service could be a reasonable goal in health-care conscientious objection, as long as that service benefits the community who may otherwise be harmed by a provider's conscientious objection.

My own contribution to the volume, "Of Mites and Men: What WWII Scabies Experiments Can Teach Us About New, Unregulated Human Subject Research," considers what lessons Mellanby's work may teach us about ethical approaches to new forms of research not subject to regulation. I argue that when researchers – no matter how well-intentioned or ethically motivated – take themselves to be the measure of appropriate treatment, they will frequently miss important ethical considerations. Research without oversight can favor too liberal interpretations of autonomy and consent and convenient interpretations of how to make efficient use of participants. Independent researchers (or "untrammeled" researchers, to use Mellanby's term) can also fail to notice that they lack appropriate expertise for the work or fail to understand accurately what constitutes true collaboration from the point of view of participants in the research. History demonstrates that good intentions are often insufficient to protect vulnerable subjects of research.

It is an honor and a pleasure to bring this important volume back into print.

Department of Philosophy
University of North Carolina, Charlotte
Charlotte, NC, USA

Lisa M. Rasmussen

Contents

Part II Commentary Essays

Part I
Reprint of Second Edition of *Human Guinea Pigs*

Preface to the 1945 Edition

This book is an account of the way in which a number of people have voluntarily allowed themselves to be exposed to diseases, discomforts, dangers and unhealthy dietetic restrictions in order to try to help in the advancement of medical science. They have become more than simple, passive guinea pigs, whose every activity has had to be carried out under close watch and supervision, for they have taken an active part, co-operating in the work and thus making possible experiments which have often been considered impracticable. The volunteers are (or at any rate were at the outset) all pacifists with conscientious objections to taking a combatant rôle in the war. I myself am not a pacifist, but for 3 years I have lived and worked with these volunteers and I think it is possible for me to give a fairly detached view of them and of the contribution they have made to research and medicine. It will appear that the volunteers, except for their views on war, were a fairly normal selection with perhaps rather more virtues and rather fewer vices than the average members of the population, but for the most part they were in no way either saints or "cissies". Some were diligent, a few were bone idle. Most of them were of more than average intelligence. But in addition to their pacifist views (and these were by no means uniform) they had one thing in common throughout the whole of the long period through which they served as human guinea pigs—they co-operated in the experimental work with complete trustworthiness and loyalty. Never in any way were the experiments "let down" by the volunteers; I think that this was a remarkable achievement on their part which deserves the highest praise. It was for this experimental work that they were engaged and if a few proved somewhat irritating in their apparent inability to get on with various domestic and other jobs which were not directly concerned with the experiments, it must be realised that the conditions of the experiments were such as to make for imperfect health. But the majority were able to co-operate with the experiments and at the same time get on with some hard work when necessary.

L. M. Rasmussen (ed.), *Human Guinea Pigs, by Kenneth Mellanby: A Reprint with Commentaries*, Philosophy and Medicine 134,
https://doi.org/10.1007/978-3-030-37697-0_1

There is a second theme to this book. It deals with the way in which certain biological experiments were planned and carried out. This is described in detail because it seems to me that it may be interesting as a piece of scientific organisation—or perhaps it would be better to say "lack of scientific organisation." In these days, when there is so much emphasis on the desirability for scientific planning, many comprehensive and rather grandiose schemes for research are being discussed. Some scientists feel that there is no longer a place for the individual carrying out research in his own particular way, and that progress can only be made by well organised teams of specialists. I realise that such teams are necessary in many fields, particularly in war time, but I personally also think that scientific progress will be sorely stunted if there is no place for the individual to develop his own particular type of work. The investigations described here contain what is essentially a personal and individual story, and while I must apologise for looming so large in it myself, this is unavoidable because the whole investigation has been such a personal one. It has been necessary, for instance, to deal in Chap. 1 with some of my personal problems and with work which does not apparently concern the human volunteers, for this was an essential preliminary which led directly to the later investigations.

Although this was so much a personal investigation, any progress it made owes much to team work, the team consisting first of the experimental subjects and secondly including several scientific workers with whom I have collaborated. But instead of working as a highly organised team the various research workers have co-operated informally on points of common interest—this informal and personal co-operation seems to me to be more likely to be fruitful than the most efficient regimentation and organisation of scientific work. I am afraid that in a community in which research is perfectly organised, schemes such as that described in this book might appear far too scatterbrained to obtain support. I sometimes fear that many lavishly financed and efficiently organised schemes will often be sadly sterile, for to my mind, in research, inspiration and organisation by no means always go hand in hand. I hope that there will always be a place, and funds, for the individual who wishes to work in his own way, untrammelled as much as possible by the "red-tape" which seems to be a necessary accompaniment of any large-scale organisation.

Preface to the 1973 Edition

The greater part of this book was written, and published, during the 1939–45 war. The book was then well received and widely reviewed, and two printings were soon exhausted. However, as paper was very scarce, further reprints were impossible and the book quickly went out of print. Readers may wonder why, nearly 30 years later, it has been decided to bring it once more out of oblivion.

The first reason is that the original book concerns a somewhat off-beat but nevertheless interesting facet of war history—a subject in which there is today a great revival of interest. The part played by the group of pacifist volunteers deserves to be remembered. Secondly, the disease of scabies (or "the itch"), a war-time scourge which they did so much to control, and which almost disappeared after the war, has now become epidemic once more. This is, in part, because the lessons learned in the research with the volunteers have been forgotten.

The third reason for wishing to revive this book is that it makes some contribution to the present debate regarding the planning of scientific research. We have recently had the revolutionary proposals for the reform of government research made by Lord Rothschild, and adopted in principle by the Government in the face of almost universal condemnation from every responsible scientist in Britain, as well as by the Parliamentary Select Committee on Science and Technology. Lord Rothschild has proposed that instead of allowing the Research Councils, which are bodies whose members come mostly from universities and are independent of government, but which are almost entirely financed by government, to continue to have full responsibility for planning and commissioning their research programmes, much of the money previously granted to them by Parliament should now be paid to government departments. These departments would then decide how, and on what type of research, this money would be spent. Under this system the department is

L. M. Rasmussen (ed.), *Human Guinea Pigs, by Kenneth Mellanby: A Reprint with Commentaries*, Philosophy and Medicine 134,
https://doi.org/10.1007/978-3-030-37697-0_2

described by Lord Rothschild as the "customer"; if a research council carries out its proposals it then becomes the "contractor". Although Lord Rothschild's proposals have been generally condemned, nevertheless most scientists agree that our mechanism for choosing the type of research to be supported with government funds could be improved. The scabies investigation deserves to be examined in this connection. Without wishing to appear conceited, I think that I can claim that it was, in practical terms, one of the most successful programmes ever initiated. It was directly responsible for administrative changes in the treatment of the disease. Whereas previously soldiers and civilians had frequently spent weeks in hospital, often suffering from secondary lesions caused by maltreatment of an already-cured condition, it was shown that a cure could be virtually ensured in under an hour, and when this discovery was implemented it was calculated that this alone freed the equivalent of two divisions of soldiers from hospital. Even greater numbers of civilians must have been set free. In addition the demonstration that there was no need to disinfect all clothing, bedding and household goods of patients has saved several million pounds each year to local authorities from that day until now. The total cost to the government of the scabies work came to under £5000. If this piece of work paid these enormous dividends, surely it should have been the ideal instance of one where a "customer", wishing to improve treatment and to save money on prevention, should have been searching the highways and byways for a "contractor". In fact nothing of the kind occurred. Every medical administrator realised there was a scabies problem, but no one in authority recognised that it was a problem suitable for this kind of study. The work was only done because a very junior research worker thought that something should be done, and then put forward a hare-brained scheme which, miraculously, attracted a modest amount of support, even if this only became operative after he had crusaded for this support with missionary zeal. Another lesson may be learned regarding the application of the research; here again I must take some credit, as I took responsibility for trying to get my results applied as widely as possible. But it says much for the authorities—in this case the medical scientists in what were then called the Ministry of Health and the Board of Education—that they did in the end agree to support the work at all. And I think I show that they, and also the medical directorate of the armed services, were prepared to be convinced by results, even in face of opposition from conservative elements in the medical profession.

Another point which may be of interest to those who are planning research policy is that the most valuable results were not those which I set out to obtain. I was originally concerned with a simple and practical problem, which was: How long will bedding or clothing remain infective without being disinfested? When I found that infection by these means was so unimportant, had I been well disciplined I would at once have discontinued the work. Had I been under the strict control of an unimaginative Director or Committee, the work would have been stopped. Instead, I was able to continue and clear up many other more important and practical, though unforeseen, points about the disease.

One other facet of the work with the volunteers may be mentioned. I soon found that the scabies work alone did not make full use of their potentialities, and so they also served as subjects in dietetic experiments. This was done at my initiative, but the scientific work became a co-operative effort of a group of distinguished investigators. The studies on vitamin A and vitamin C were of a kind never done before, and unlikely to be attempted again, as they consisted of studies of humans living for several years on controlled, deficient diets. The Medical Research Council reports based on these experiments are still classics in their field. I think all scientists should take note of this example, and try to make as efficient use of all their material.

Artistically it might have been best to have entirely rewritten *Human Guinea Pigs*. I decided not to do so, as the original account, naïve as it may be in parts, is a contemporary account of the work, and so it has an otherwise unobtainable air of authenticity. But I have added some additional material. Thus I give a brief account of the other experiments performed on my volunteers after I joined the army in 1943 and went to North Africa, South-East Asia and New Guinea. The vitamin A experiment went on until the late summer of 1944, and the even more testing experiment, on vitamin C deprivation, continued until February 1946. After this the Sorby Research Institute came to an end, although many of the volunteers kept in close touch with each other, through a newsletter and by other means, for several years. In 1972 the BBC made a television programme which included their work on scabies, and we considered using some of the genuine guinea pigs themselves. However, the producer preferred to use young actors, as the original participants were, and looked, 30 years older than when they were the subjects of the original work. I myself made two brief appearances in the programme, once photographed specially in 1972, the other in the teaching film made in 1942. I do not think that the viewers all realised that it was the same person, though I received a number of letters from people I had known many years before during the 1940–43 experiments and who recognised the older pictures.

Other additions deal briefly with other successful experiments using volunteers. During the first months when I was in the army, I organised a supply of volunteers from the Friends Ambulance Unit to serve as subjects for malaria work under the direction of Professor Brian Maegraith, who was in charge of malaria research at the War Office. He also had made trials of the toxicity of anti-malarial drugs on himself and his colleagues, and on some 480 Oxford undergraduates. One of the most brilliant pieces of work on malaria was that done by Sir Neil Hamilton Fairley, using Australian soldiers as experimental subjects. This directly resulted in altering the whole course of the war in New Guinea and the South-West Pacific, by reducing the incidence of malaria from unmanageable to almost negligible proportions. Finally I mention the work of the Common Cold Unit at Salisbury, a project partially foretold in the last chapter of my original book. The main addition, however, is the section on the experiments in the German Concentration camps. If the other sections describe research done in a reasonably ethical manner, this chapter indicates the dangers of such investigations, if they are done in a place, and under a political system, where civilised standards are not maintained.

It will be seen that this book deals only with medical experiments on otherwise healthy subjects, not with the vexed question of trials of new methods of cure on actual patients. This subject has often been dealt with elsewhere, and an approved code of practice has been laid down by the Medical Research Council and other authorities. Nevertheless, there is still some concern lest academically minded clinicians may sacrifice their patients in the hope of improving the standard of treatment in the future. This view was expressed somewhat emotionally by M. H. Pappworth in a book he published in 1967 with exactly the same title as mine, i.e. *Human Guinea Pigs*. I must admit to some irritation at this plagiarism on his part as although my original book was out of print there are copies in most libraries and it appears in many bibliographies. Also I was surprised to find no mention of most of the important human experiments even when these have been fully described in the medical literature. He might be excused for omitting my scabies work on the grounds that scabies seldom kills, but he seemed unaware of the vitamin trials, which were obviously important, classical experiments. Even more surprising he ignored all the malarial work, even that done in Australia by Fairley, and he did not even mention the common cold trials. A comprehensive work on this subject is still required.

There are clearly still many gaps in medical knowledge. Progress in chemotherapy is being delayed because, in the hope of making sure that every new drug is completely safe, almost impossibly stringent standards are being insisted upon. I think that we still need facilities by which volunteers anxious to promote medical progress can be allowed to give their services, even if there is some degree of risk to their own health or even, in extreme cases, to their lives. We encourage mountaineers and yachtsmen to take calculated risks in the name of "sport". I am sure that many more would welcome the opportunity to take similar risks for the benefit of all mankind. However, they must be genuine volunteers. I do not think, particularly in view of what happened in war-time Germany, that any type of prisoner should ever be used.

Chapter 1
Biological Research in War-Time

Curiously enough, and perhaps due to an upbringing which included four years in a junior and three years in a senior Officers' Training Corps, I had always imagined that if and when war came I would find myself immediately in the Army. Actually, to me, as to many others, what happened in September 1939 came very much as an anti-climax. I had spent the last days of peace in a somewhat unreal and theatrical atmosphere. After the signing of the Russo-German Pact on 23rd August, I was the only British delegate to turn up at an International Scientific Conference in Utrecht in Holland, where I was therefore given an undeservedly warm welcome and where my paper on "Rhythmic Activity in the Bed Bug and the Cockroach" was received with a degree of interest and appreciation which were not entirely deserved. The Conference ended with an excellent party in a mediaeval Dutch castle, with Dutch hospitality at its most alcoholic, after which I caught the Rheingold Express and the last boat to leave the Hook for Harwich before the outbreak of war. In the train an anti-Nazi German-American, whose relatives had apparently already suffered in Germany, and who believed that war would begin at any moment, insisted that I should help him in wrecking as much enemy property as possible, so that the windows and luxurious fabric of the Rheingold first-class carriages suffered accordingly. I hope that the Dutch, in whose territory this damage was done, were not made to pay for its repair. Somehow or other I had obtained a cabin on the boat, which was carrying several times its normal complement of passengers, so that sleeping bodies covered the corridors several deep.

As soon as I returned home I found myself faced with the task of trying to get a Latvian boy of 16 back to his native country. This was accomplished after an all-night train journey in a blacked-out train full to overflowing with semi-drunk fisher girls returning to Scotland. I was able to arrange for a sea passage for the boy to

L. M. Rasmussen (ed.), *Human Guinea Pigs, by Kenneth Mellanby: A Reprint with Commentaries*, Philosophy and Medicine 134,
https://doi.org/10.1007/978-3-030-37697-0_3

Oslo and he reached his home eventually after a train journey from Oslo to Stockholm, a boat from Stockholm to Tallinn in Estonia and home from there by train. He arrived home safely but since the Russian occupation of his country in 1940 I have not heard any more of his family.

After these experiences it seemed natural to expect to find oneself playing an active, if not an heroic, rôle. What actually did happen was that, thanks to the Schedule of Reserved Occupations, I found myself effectively barred from becoming a combatant. The Authorities, no doubt wisely, wished to avoid the mistakes of the last war, when so many men, whose professional or scientific qualifications would later have been valuable, joined the Army in 1914, so that the country was deprived of their special abilities. In order to make the best use of all technically-qualified men and women in any emergency, the "Central Register" was set up some time before the outbreak of war. A series of elaborate cards was sent out to everyone with any scientific qualifications, and we were all given the impression that there were lots of important technical jobs which we might at any moment be summoned to fill. I believe that on the whole the Central Register has done invaluable work, particularly in respect to physicists and chemists, but as regards biologists the position has not been so satisfactory. In September 1939 I enquired of the Central Register if there was anything they would like me to do, and was informed that at the moment my services were not required. After that I received no communication for two years, when it was suggested that I should appear for an interview to be trained for some work in connection with radio location, work which could equally well be carried out by almost any undergraduate. I know of other biologists who filled in their forms religiously and who have received no communication whatever for four years. This apparent waste of manpower is not altogether the fault of those running the Central Register. There are many peace-time professions which have little value in war-time, and the average biologist is not, in fact, particularly suited for technical work aimed at slaughtering one's enemies. But if the biologist could not be used in a technical capacity it appeared unnecessary that he should be prevented from becoming a combatant because of his "reservation." However, though we biologists may think otherwise, we are only a very small and unimportant section of the population and it is not surprising that we found ourselves in the position of "forgotten men."

This situation caused a good deal of concern to some of my biological colleagues, who felt that the authorities should be able to find them at once work of national importance. I think that many of the criticisms of the authorities were unjustified, for we biologists were the experts in our own subject and therefore, instead of simply asking to be used in some way, we should have used our expert knowledge to suggest actual practical problems which we were fitted to investigate. As will appear from this book, suggestions of this kind often received the greatest possible support, though there are, unfortunately, other occasions on which valuable projects have received insufficient consideration.

My position, therefore, at the outbreak of war was that I was not allowed to join the Forces, except in some technical capacity, and all offers of my services, though made through what I thought were most influential channels, received little encouragement. Furthermore, nobody could suggest any problems on which I ought to work in order to help the national effort. Some explanation of my own position is here necessary. At the outbreak of war I held the Sorby Research Fellowship of the Royal Society. The funds for this Fellowship were left by the late Dr. H. C. Sorby, F.R.S., with the proviso that the holder should, if possible, carry out his investigations in the University of Sheffield. (Dr. Sorby had himself lived in Sheffield, carrying out his own important research there, and making many important contributions towards the University.) For ten years up to the outbreak of war I had been engaged in research on various problems concerning insects, both in this country and abroad, and as Sorby Fellow I had continued these investigations. The conditions of my tenure were somewhat elastic, and the Committee of the Royal Society who were responsible for managing the Fellowship were prepared to allow me to carry out some biological research which was of practical importance in war-time, instead of continuing strictly to follow my more academic programme. The first task was to locate a problem, the solution of which would be of immediate value. In any really long-term programme of research, progress either academic or of direct practical importance is only likely to be achieved by taking a broad view of the subject, but in war-time workers hardly feel encouraged to continue investigations which are unlikely to yield results for a very long time, and prefer to tackle simpler problems of more immediate value. Incidentally, it is interesting to note that, whereas in the past one has usually expected to be able to get practical dividends eventually from purely academic studies, much work that has been done during this war and is directed to immediate practical ends has, at the same time, given us fundamental results of a kind hardly expected to be obtained from such programmes.

For some time, Professor P. A. Buxton, of the London School of Hygiene and Tropical Medicine, remembering what an important problem had been produced by the louse in the last war, had been carrying on investigations which have been very successful in improving our methods of coping with this insect. In October 1939 he allowed me to try out, on a small scale, certain preparations which were being tested at that time. I was able to make arrangements to carry out these tests among patients admitted to a local children's hospital. When I started this investigation I had, like so many other people, imagined that the head louse, which had been almost universal a couple of hundred years ago, was now a rare creature and only to be found among the destitute and uncared for. I had seen reports of School Medical Officers which deplored the louse's existence among their children, but many of them gave figures which suggested that the incidence was very small so that they did not present an important problem. On the other hand, reports in the newspapers alleged that the children evacuated from our cities at the beginning of the war were found, very commonly, to be verminous. I am afraid I had taken these newspaper reports with a pinch of salt, and assumed that "blimpish" country dwellers in a hitherto protected environment had greatly exaggerated a very unimportant problem which had been thrust upon their notice. When, however, I came to examine children recently taken into hospital I soon came to the conclusion that any misrepresentation on the part of the newspapers at the time of evacuation had been to minimise the problem. I was appalled at the number of children with lousy heads.

As a result of a small investigation on head lice I made a number of statements which sharply criticised the publications and policy of the Board of Education and its health services. I am happy to say that the result was not that the Board refused to face these charges, but instead I was given all possible encouragement, together with an expenses grant from the Board and the Ministry of Health, to investigate the whole matter thoroughly. One frequently finds public servants who resent the slightest breath of criticism, so it is therefore a great pleasure to be able to state that Dr. J. Alison Glover, who was the Senior Medical Officer to the Board of Education, did everything in his power to make the louse investigation a success, though my comments on the Board's health services had been very far from tactful, and though the whole study was, in effect, a criticism of the School Medical Service.

This investigation on the incidence of head lice was begun before Christmas of 1939, and the report, which summarised the results of some 60,000 examinations, was submitted during the summer of 1940. The general result was that, though the children in rural areas were generally free from vermin, head lice were distressingly common in our cities, where as many as 50% of the girls of school age might be affected. The results of this work have been published in full elsewhere and incorporated in circulars issued by the Board of Education and the Ministry of Health (No. 1544 Board of Education, No. 2306 Ministry of Health); they are also discussed in a book called "Our Towns" by the Women's Group on Public Welfare (Oxford University Press, 1943).

While working on head lice I came into contact with a great number of Medical Officers and others who are concerned with public health, and by talking with them I soon realised that, although lice presented a problem, the disease called scabies or "the itch" was, if anything, a more serious matter. This disease is caused by the itch mite, a tiny eight-legged creature which burrows under the skin, and there were many problems concerning scabies which could best be tackled by an entomologist. The question was, how could such an investigation be carried out?

Chapter 2
Scabies as a Research Problem

Scabies was obviously a problem which it was worth while tackling, and it was a problem best tackled by one with an entomological background. The problem was, how to set about tackling it?

It is clearly necessary at this point to include a few general comments about this disease, and the state of our knowledge in 1940 when this investigation was being planned. Scabies is defined as "an intensely itchy disease of the skin caused by the presence of the itch mite *Sarcoptes*." This little creature, properly called *Sarcoptes scabiei* (variety *hominis*) is a mite or acarus (both these words "mite" and "acarus" are general terms and can be used to refer to a *Sarcoptes* or a cheese mite, and to a whole host of other little creatures in the same way as "insect" is used to refer to a bug or a beetle or a bee or a butterfly). Most mites are tiny free-living organisms, subsisting on a diet of vegetable juices, but some suck blood and a few are parasitic on higher forms of life. The itch mite is parasitic, and *Sarcoptes scabiei* of the variety *hominis* attacks man. This acarus is about a sixtieth of an inch in length, and it lives in the human skin, tunnelling about in the outer horney layer or *stratum corneum*. The presence of the mites sets up intense irritation, hence the common name for scabies, "the itch"; as a result of the irritation the patient scratches and this may damage the skin and allow it to be invaded by other organisms so that local septic infection is apt to occur. Scabies is cured when all the invading parasites are killed. There are many substances which will do this, but unfortunately the human skin is a delicate organ and the difficulty is to exterminate the mite without seriously affecting the skin. Scabies is transmitted by the transference of the parasitic *Sarcoptes* from one person to another.

L. M. Rasmussen (ed.), *Human Guinea Pigs, by Kenneth Mellanby: A Reprint with Commentaries*, Philosophy and Medicine 134,
https://doi.org/10.1007/978-3-030-37697-0_4

Reference to the medical press of 1940 will reveal how great was the chaos regarding the treatment and prevention of scabies. Various authorities advocated different remedies and treatments, and insisted that the other's remedies and treatments were less satisfactory. There was disagreement on the way in which the disease was transmitted, and on the risks of picking it up under various circumstances. A large proportion of patients was being submitted to treatments over periods of weeks, months and even years without relief. Some children grew up illiterate as they had scabies at five years old and were not cured by the time they were 14. Thus they were entirely excluded from school and had no education whatever!

It is interesting and salutary to note that, though this gloomy picture is a true one, the situation need not have been like that. Among the mass of writing about scabies there were to be found lucid accounts of first-class work, but unless one had made a particular study of the subject it was almost impossible to sift these from a mass of indifferent and inaccurate papers. Thus the great Viennese dermatologist Hebra discovered almost all the main facts about scabies a hundred years ago, and if proper reliance had been placed on his results, instead of their being practically forgotten and displaced by inaccurate information based often on the study of unrelated parasites attacking animals and not man, a great deal of human suffering would have been avoided. Almost the only dermatologist in this country to stress the importance of Hebra's work was Dr. A. M. H. Gray, and I shall always be grateful to him for insisting that I should study these older papers. Then regarding treatment: the Danish dermatologist Kissmeyer a few years before the war described a simple method which gave excellent results. Even this method was not entirely new; a similar treatment had been advocated with success nearly fifty years ago, and if it had been generally adopted (as it was with success by some) these cases of so-called chronic scabies causing months of suffering and discomfort would have been avoided. Unfortunately many discounted his results without proper trial and then themselves made use of "remedies" which were sometimes worse than the disease. To a great extent my investigations were to show that earlier workers had already solved many of the outstanding problems regarding scabies, but that these solutions had often been overlooked.

There were obviously two possible lines of attack in any programme concerning scabies. The first was the clinical one, i.e. mainly to study the disease in man from the point of view primarily of a human disease. The second was more entomological and entailed a detailed study of the parasite itself in the belief that by an understanding of the cause of the disease we would be in a better position to effect its cure and prevention. The entomological approach appeared to me to be the logical one, and also it was the one which I was qualified to make. For such an investigation I would require access to parasites and laboratory accommodation for the experiments.

The scheme of research as it developed eventually only slowly took its final shape. I must make it clear that at the outset the problem which appeared to me of paramount importance was that of transmission, that is to say, just how and where did the uninfected person pick up the scabies mite and so "catch" the disease. I believed, with most others, that infection usually took place through the medium of inanimate objects, i.e. that mites were to be found lurking in dirty blankets waiting to infect the unfortunate individual who used them next. I was even prepared to believe that the scabies epidemic which was apparent in these early days of the war was mainly due to the military, for soldiers were, I thought, being infected by the bedding which was not being frequently enough disinfested, and so they were then taking home the disease to infect their wives and children when on leave. It was only later that this was seen to be quite the reverse of true. I particularly wanted to know how long the parasites would survive away from man, for this information would tell us just how long a billet must be left empty, or bedding must be left in store before it could be relied on not to transmit scabies to the next user. It will be apparent that the investigation soon took on quite a different direction from this.

To work on any biological problem, the first requisite is a supply of biological material—in this case a supply of *Sarcoptes*. Several courses were possible to obtain this material. It might be possible to obtain parasites from animals—for instance, the horse has its own form of scabies which is caused by a *Sarcoptes* very like that of man; as will be described below, attempts to obtain such a horse were made but proved fruitless. Other animals have different forms of mites parasitic on them, and some of these are not unlike the organism causing human scabies; these could have been used but I felt that it would be preferable to use the actual creature which caused the human disease. This meant getting access to patients with scabies and removing the living parasites, or else having volunteers willing to submit to infection and serve as "breeding grounds" for the mites; or (as I then thought) we could rely on the bedding and perhaps the underclothing of persons infected with the disease. But however the problem was to be tackled, facilities, support and funds from some appropriate body were obviously required.

During the early summer of 1940 I gave a good deal of thought to these problems. I attempted to collect preliminary data and took every opportunity to discuss the matter with those people who had some practical experience of scabies. During this period I was carrying out the head-louse survey mentioned in the preceding chapter, and several times I was able to discuss scabies also with Dr. Glover. I informed him that I was becoming interested in further research on scabies—he was one of those who had often referred to the problems it caused among school children—and it was at his suggestion that I took the next step which was eventually responsible for the whole investigation.

Sir Weldon Dalrymple-Champneys, the Deputy Medical Officer of the Ministry of Health, was also concerned with an inter-Service committee which dealt with the problem of scabies in the Forces; also the Ministry had wide responsibilities for advising on the treatment of scabies among the civil population. On August 15th 1940 I wrote to Sir Weldon saying that I had in mind a scheme for further investigations on scabies, and on receiving an immediate and favourably worded reply I sent the following rough draft of a scheme for the proposed work:

Scabies Investigation

Although the diagnosis and treatment of scabies appears to be comparatively simple, yet infestation is all too common among the civilian population, and in the Fighting Services these parasites appear unpleasantly frequently. It appears likely that further knowledge, particularly of the biology of the mite, would be valuable and would go far to help to eradicate the pest. The following are suggestions for further research with a view to gaining such knowledge:

(1) Laboratory Experiments

Accurate information is required about the life history of the mite, particularly regarding the length of survival of the infecting stages away from the host. It will be necessary to consider the effects of climatic factors, particularly temperature and humidity. The information gained should enable us to say exactly how infection occurs, and how long fomites are likely to remain infective at different seasons (a point of considerable importance in billeting, etc., when disinfestation is impossible). Further knowledge will also certainly show how destruction of this parasite may be more easily accomplished.

Experiments of the kind referred to above will not be easy to perform, due to the small size of the mite and the difficulty of obtaining it in large numbers. Co-operation with the Army Medical Staff, and funds sufficient to procure technical assistance, would facilitate such work very considerably. As things are at present I shall be able to carry on such work on a small scale, perhaps assisted by various local public health authorities with whom I have contacts, but if the investigation could be carried out on a larger scale, more valuable results would be obtained more quickly. It is possible that mites of the variety which attacks the horse, and which can be obtained in large numbers, would be useful in doing the preliminary experiments.

(2) Field" Work

Little is known about the transmission of the mite. A rather small number of experiments, using human volunteers, has given some information, but much more needs to be done on these lines. This could be carried out in a military hospital (as was the previous work) but it would be much more satisfactory to set up an institution specially for this work. If the work were done on a large scale this would also be more economical, as the overhead running expenses would be small. The details of any such scheme require careful consideration.

It might also be valuable to collect and correlate hospital and public health data concerning scabies in several areas and compare the treatments given. The procedure is far from uniform and in many districts could be greatly improved. It appears probable that some foci of infestation in the country would be revealed and eradication of these would simplify the whole problem."

It may be noted that in the document quoted above reference to "human volunteers" occurs. The idea of using Conscientious Objectors had been growing in my mind for some time, and I knew that at this time there were many individuals who felt that they could not take part in the war as combatants and who at the same time wished to serve humanity. In the summer of 1940 the number of outlets for such service was small, and many pacifists appeared to think that the duties which they were performing, or to which they had been directed, were very unimportant. Preliminary discussions had suggested that there would be a good response to an appeal for volunteers to serve as subjects for medical research, provided that it could be shown that this would be likely to be of value in alleviating suffering, and provided that it was not solely directed to improving military efficiency (most pacifists realised that this type of work in war time was bound to be made use of in the forces, they did not wish to take part in experiments solely directed to that end).

After my proposals had been acknowledged by the Ministry of Health there seemed at the time to be an interminable and unnecessary delay before they were put into practice. Looking back at what actually happened in the light of later experience I am more surprised at the speed at which everything was managed, for the Ministry was committing itself to a type of programme quite unlike anything it or anyone else had tackled previously, and there were many people whom they felt they had to consult in addition to the usual difficulty of obtaining sanction for the expenditure from the Treasury. I do not think that "red tape" was allowed to cause any real delay, and as these discussions were being carried out when the Ministry and most people in London were greatly preoccupied by German air raids, I think the way things developed reflects considerable credit on the oft-maligned civil servant.

On October 10th 1940 I attended a conference at the Ministry of Health with Sir Weldon Dalrymple-Champneys and Professor Buxton. It was indicated that the Ministry was favourably inclined to the whole proposal, all aspects of which were discussed, and I was then asked to put forward a more detailed programme which would include financial estimates for the various parts of the investigation. The following is an extract from the proposals which I submitted on October 16th:—

Scheme for Scabies Investigation

This scheme is drawn up assuming the investigation would take one year.

Laboratory Experiments on the Biology of the Parasite (Sarcoptes scabiei, var. *hominis)*

Experiments will be made to determine the longevity and resistance of all stages under a variety of conditions of temperature and humidity. Practically no experiments have been made since 1918. Laboratory technique is so much improved that it should be possible to obtain the results required within a year. Some work of this nature is being begun in a preliminary way. Arrangements have been made to obtain material from several hospitals and the cleansing station in Sheffield but it is hoped to arrange to obtain much of the material from the military authorities. It should be possible to do some preliminary experiments using the abundant material obtainable from an infested horse which is attacked by a biological race of *Sarcoptes scabiei*, and preliminary arrangements for the accommodation of such an animal have been made. Some assistance for this work may be given by members of the staff of Sheffield University, but it will be necessary to engage a full-time laboratory assistant and additional clerical assistance will be required. Most of the apparatus required is already available.

Expenses: Assistants' salaries	£220
Laboratory expenses and local travelling	50
Maintenance of horse	20
	£290

Experiments Concerning Transmission of Scabies Using Volunteers

To obtain proper information on transmission experiments must be made using volunteers and it will be most satisfactory if these men are under my control. Preliminary exploration shows that it may be possible to obtain suitable premises for accommodation of volunteers and for the pursuit of these experiments from the city medical department. If this should prove impossible, other convenient premises are available and can be rapidly and cheaply adapted. It appears probable that a number of men registered as Conscientious Objectors will be willing to co-operate and it is suggested they be given pay, etc., similar to that received by a private soldier. The most suitable of these men would be employed to help in the laboratory and the others could best be kept busy by taking over some nearby allotments.

Experiments with volunteers will first be directed towards finding how long fomites, etc., remain infective. This work will also help to check the laboratory investigations, and enable us to see not only how long a given parasite will *survive* but also how long it will remain *able to infect*.

All physicians of my acquaintance agree that diagnosis is often unsatisfactory and that many cases described as scabies are, in fact, not infected with *Sarcoptes*. It is obvious also that many true manifestations are never recognised as such. Careful examination of controlled infections should make it possible to produce a more accurate description of the disease.

There is little agreement as to the best treatment of scabies and none of the methods at present employed appears entirely satisfactory. It will be possible to compare results of various methods used in several institutions (all these institutions use different treatments) and find out exactly how each treatment affects the parasite. If this is done in collaboration with the various physicians, it should be possible to decide which method is really the best and probably to introduce further improvements.

It is suggested that about 12 volunteers be employed for a period of about a year, and it is difficult to forecast the exact cost, but enquiries concerning expenses at the University Hostels and the experience of billeting officers suggest that this should be covered by the sum of £1450. If adjustments in the number of volunteers can be made when necessary, it should not be difficult to keep within this figure.

Co-operation with the Services

I wish to co-operate wherever possible with the medical staff of the Navy, Army and Air Force. The Army authorities in this neighbourhood will be able to supply much material. I wish to meet the officers concerned in all the Services as soon as possible to discuss problems with them and to find how we can co-operate further. It is probable that such co-operation would mean that arrangements could be made for the practical application of results of the investigation to be tried out on a large scale in the Services without any undue delay.

To keep in touch with the Services' Medical Departments and Health Departments in other cities, a good deal of travelling may be necessary. It would be convenient to have available a sum of £100 for travelling expenses and subsistence allowance.

These proposals were in two parts for this reason. The whole scheme was more or less my "ideal plan" for the investigation. The first part, comprising the laboratory experiments, I felt pretty sure would be agreed to. On the other hand, although to me the proposal to recruit pacifist volunteers and to run a special research institute for their study seemed a good one, I was not too sanguine that it would appear practicable to the Ministry or to the Treasury. However, on November 29th I heard from the Ministry of Health that my detailed proposal had been agreed to and that the funds requested would be available. I still felt that this revealed a surprisingly progressive attitude for a Government Department and, if it does not appear patronising, I would wish to accord the Ministry full marks.

Between the middle of October, when the detailed plans were submitted to the Ministry, and the end of November, when they agreed to give their support, I had been getting ahead as rapidly as I dared. I had no definite promise of support, but sufficient encouragement had been given to make me feel that it was worth while getting everything ready. I was faced with the problems of finding somewhere for the volunteers to live and work, and also of finding the volunteers themselves. The most difficult thing was to time everything correctly—I did not know when, or even if, the money would be available, but unless I could produce accommodation and volunteers simultaneously then I would have put the Ministry to a great deal of unnecessary trouble, and I and the Ministry's scientific advisers, who had said that the project was a possible one, would be in an invidious position.

The housing question was not a difficult one. In the earlier part of the war there were many large empty houses in all cities, and I soon saw a number of suitable premises. Furnishing and equipping were more difficult, but there was as yet no shortage of furniture or soft goods, and prices obtained at auction sales for second-hand furniture were ludicrously small when compared with those of today.

There are several Pacifist organisations concerned with social service, and I got in touch with these in order to recruit my volunteers. The greatest possible assistance was obtained from the Pacifist Service Units; the Headquarters Staff were sympathetic to the scheme, and were responsible for finding the majority of the subjects. This organisation took a refreshingly realistic attitude to their members, and made no pretence that just because a man called himself a Conscientious Objector he was therefore a paragon of all the virtues—in fact they warned me that I was taking on a very difficult set of human problems. It says much for the care and skill with which they selected the volunteers that so few problems of the kind we had visualised did in fact arise.

Another organisation with which I made contact was the Friends' Ambulance Unit. They were also sympathetic, but their organisation is such that members are normally employed in sections and so it was not possible for them to detach odd members for a job of this kind. They did suggest supplying a special section to comprise all the subjects used in the research, but at this time we had already engaged the majority of men required. In later experiments I have had very fruitful collaboration with F.A.U., but circumstances prevented our co-operation in the scabies work.

A further number of prospective volunteers was discovered among pacifists already resident in Sheffield. I met a body of these men and explained the scheme of research; several volunteered and took part in the various investigations. Altogether, particularly in 1940 when the "man-power" situation was not yet acute, there was no difficulty in finding a suitable number of volunteers; at this stage my main worry was whether they would be possible to live and work with!

Chapter 3
Preparations for the Investigation

On the first of December 1940 we took over "the house known as Fairholme, being number 18, Oakholme Road, in the Sheffield suburb of Broomhill." This house is a largish Victorian villa in a smallish garden. The American journalist who described it as a "rambling old mansion" relied more upon his imagination than upon his powers of observation. Although the built-up part of Sheffield stretches out for another couple of miles, this district is well supplied with trees and contains a number of houses with grounds of many acres and even some open fields. The house had not been long vacant and was in excellent repair—so many of the empty houses which we inspected would have required extensive decoration and repairs before they would have been habitable.

Ideally a laboratory should always be built as such, and should be planned in consultation with the person who is going to work there. Even when a building has been built as a laboratory, however, it may be unsuitable for its purpose, and therefore this dwelling house which was built for an upper-middleclass business man, his family and his domestic servants to five in was by no means as inconvenient a place for research as might have been anticipated. The rooms were large and light and one always had a sense of space which could never be afforded in even the best planned modern laboratory. I am by no means sure that I would not rather adapt a dwelling house for scientific work than use any but the most recent laboratory planned by someone else and obsolescent. Our Universities are largely constructed of massive buildings which cannot be conveniently used for any purpose whatever, and which cannot be at all easily adapted to the requirements of modern scientific workers. Expensive buildings for scientific research are too often a waste of money.

In under a fortnight the empty house was completely fitted up with the furniture and furnishings required by a dozen or more residents and with the scientific equipment required for the laboratory work. My secretary, Mrs. J. R. LaCroix, and I had

L. M. Rasmussen (ed.), *Human Guinea Pigs, by Kenneth Mellanby: A Reprint with Commentaries*, Philosophy and Medicine 134,
https://doi.org/10.1007/978-3-030-37697-0_5

already prepared provisional lists of goods required and we just went out and had an orgy of buying. The plan was to sell everything for what it would fetch at the end of the experiment.

To obtain bedding, towels and such-like was simple—we simply purchased the requisite number of dozen of each from whatever shop seemed willing to offer a reliable article at the cheapest price. Furniture was more difficult to come by. We haunted the second-hand sale room and I believe that the local furniture merchants decided that we had gone into the business in a rather humble way. Attendance at about half a dozen sales was sufficient to provide practically all the furniture we required. This included mainly tables, chairs, beds, chest of drawers and cupboards. The total expenditure on furniture, bedding, kitchen and household equipment was only about £300; this accommodated up to 20 people, so I think we can be said to have been reasonably economical.

Having got the place fitted up in this short time, our plan was to install the volunteers and start work immediately. On December 12th my assistant, Walter Bartley, arrived from Brighton, where he had been employed at the Technical College but had been discharged from his post by the enlightened (sic) town council, due to his pacifist views. Brighton had had a good deal of enemy aerial activity, and we told Bartley that he had come somewhere quiet for a change—up to this date Sheffield had not suffered from anything in the nature of a real air raid.

At 7 p.m. on the evening of the same day, December 12th, the air raid warning sounded. There was some noise, mainly of guns, but we took very little notice until about 8 p.m., when bombs started falling uncomfortably close and we retired to our shelter. Then the Sheffield "blitz" started, and if anyone ever dares to tell a Sheffielder that any town in Britain has ever had anything at all comparable he is in for trouble. The raid went on all night and completely destroyed half the shopping centre of the city; there was not a single building within about a two mile radius of the centre which did not at least have some windows broken. It was really quite a severe raid.

The building which we had taken as the Institute got off rather lightly; about half the windows, including many of genuine plate glass which produced hundredweights of débris, were shattered, but there seemed no serious structural damage. Water (which for safety's sake the public health authorities made almost undrinkable with chlorine for the next year or so) and electricity were not cut off, but the gas supply was interrupted for a period of about two months.

Bartley, whom we had told that Sheffield would be quiet and restful after the front line existence of Brighton, had a lively reception to the city. The roof of the hotel bedroom which he had been occupying had been blown off. Also I had engaged a Mr. Roth, an Austrian refugee who formerly owned a restaurant in Vienna, to look after the housekeeping and cooking for the volunteers; during this night, which he and his family had spent in the cellar, their house and all their belongings had been completely destroyed. Bartley and the Roth family all moved in to the Institute.

At first I was very doubtful whether it would be wise to try to start work at all. I believed at that time that scabies was a very infectious disease, and that proper isolation and segregation would be impossible if there were frequent heavy air raids. Without proper control of the volunteers and of the conditions under which they lived, I thought that the experiments would be meaningless. However, we decided to do our best to overcome these difficulties. One of the large cellars had been made into quite a good air-raid shelter, with a reinforced ceiling and an emergency exit. Here we set to work and constructed a series of bunks, one for each prospective occupant of the house. The plan was for each bunk to have a number which corresponded with a bed in the dormitory upstairs, and for all the bedding to be numbered in the same way. On the sounding of an alert each volunteer was to "take up his bed and walk" to the appropriate bunk in a place of greater safety. It was quite a good scheme and gave us a lot of work, but in fact the bunks were never used. Incidentally, this demonstrated the amount of disruption of normal activities and the amount of unnecessary work in maintaining a state of preparedness that even infrequent air raids can cause.

Although in the end the shelter and its bunks were not made use of, at the time I did not feel that it would be advisable to admit volunteers until these and other special arrangements had all been made, so that start of the experiments was held up until the end of January 1941. In the meantime we prepared for the start and at the same time made an attempt to get down to various scabies problems.

The "blitz" had done a great deal of damage to Sheffield, and had wrecked the cleansing centre where, among other things, cases of scabies were treated by the city Health Department. I imagined that the overcrowding following the air raids would make the scabies problem in the city acute, and so I offered our premises to the Medical Officer of Health as a temporary and emergency treatment centre. We had all the facilities which one imagined were necessary for a scabies centre—baths, hot water and waiting rooms. Also, in preparation for the experimental work, we had installed a large heated cupboard for disinfesting clothing and bedding. I should like to draw attention to the installation of this disinfestor at this stage of the proceedings, because it has sometimes been suggested that my later view, that sterilisation of clothing and bedding of scabies cases is unnecessary, is based on some preconceived notion. On the contrary, at the outset I fully accepted the general opinion that sterilisation was necessary, and even went so far as to provide the appropriate facilities.

Our offer of premises was accepted by the Medical Officer of Health of Sheffield, Dr. J. Rennie, and some of the staff from his cleansing centre set to work in the Institute. So for several weeks we had the opportunity of studying one aspect of public medicine at close quarters.

The Health Department undertook to treat cases of scabies which were reported to them whenever the patients expressed themselves willing to accept treatment. No compulsion could be exercised because scabies was not a notifiable disease. Most of these cases were originally found at inspections of children by the School Medical Service. In Sheffield, as in some other cities, and as in most country districts, the Health Department and the School Medical Service are quite separate from one another and responsible to different committees of the Council. I believe that this system, even with the best will in the world, is likely by its complication to prejudice the efficient treatment of both children and adults; for health matters in general and for scabies in particular the family is the unit which requires treatment, and so surely all members of the family should be cared for by one comprehensive organisation. As it is, a baby is looked after by the Maternity and Child Welfare Service, which is under the Health Department. On reaching the ripe age of five the School Medical Service takes over. On leaving school the Health Department again becomes responsible, except that in the factory where the child works there is a Factory Medical Service which is responsible to the Ministry of Labour. And in some areas, if all this complication should prove too much for the mental capacity of the child and it becomes insane, as a lunatic it may be under the care of yet another Authority.

Originally, the School Medical Service treated children with scabies, but this was found to be unsatisfactory because they then became reinfected at home and the whole process had to be repeated. The later procedure was to exclude the child from school and not to re-admit it until it was free from the complaint, having been treated either by a private doctor or by the Health Department. In order to facilitate the treatment, the names of children with scabies were submitted to the Health Department and then a health visitor was sent to the home to try and persuade the whole family to accept treatment at one and the same time, so as to obviate the risk of re-infection. The system sometimes worked.

Our emergency treatment centre was not in any way responsible for this complicated administration; our duty was merely to treat cases sent to us by the Chief Health Visitor. Each day we were informed of the appointments for treatment which had been made, and sometimes the patients turned up. The health visitors did their best to persuade whole families to come, but too often some member who thought he was not infected, but whom we now know was simply incubating the disease, failed to appear. The treatment given was crude but surprisingly effective. A male and a female orderly were employed (we soon took over the treatment of the male patients) to treat the different sexes. The patients were smeared all over and rubbed well with plenty of soft soap. They were then put into a bath as hot as could be borne and soaked for twenty minutes, after which the body was roughly scrubbed with a brush "to open the burrows". Then, after drying, a liberal application of sulphur ointment was made to the body surface, and it was rubbed well in. A sufficient surplus of ointment was used to ensure that the underclothes were well impregnated and soiled with the medicament. I have myself submitted to this treatment and have seldom experienced anything so unpleasant, but it certainly gave good results. Many of the patients were children who had been neglected and were covered with septic sores—a week after the scabies treatment most of these had usually healed. While the patients were being bathed their clothes were disinfested and the whole visit lasted just about an hour. The orderlies were zealous and placed great faith in the use of the scrubbing brush. They had no knowledge of the cause of scabies, and had probably never heard of the itch mite which was responsible for the trouble, but their vigour and plenty of sulphur ointment certainly did the trick. If everyone in the city with scabies had been willing to submit to their tender mercies the epidemic would have disappeared immediately.

We certainly learned a great deal during the period in which the house served as an emergency scabies treatment centre. We had the opportunity of getting used to seeing bad cases of neglected scabies, and of learning how effective sulphur ointment (which is one of the oldest remedies) could be in their treatment. When the Health Department was able to open a new centre at one of their hospitals we were quite sorry to see their staff depart, but this happened at an opportune moment because we were now ready to receive our volunteers, and they were ready to arrive.

Chapter 4
The Human Experiment Begins

At last everything was ready and the first volunteers were expected. I arranged to take them on a few at a time, as this gradual method seemed likely to allow us to get adjusted to one another's peculiarities most easily. I must confess that at this late date I was a bit nervous about the whole business. I had very glibly said that this type of human experimentation would be possible, and might give us satisfactory answers to a lot of outstanding questions, but was it really going to be practicable? Were the volunteers going to co-operate, could we rely upon them to carry out their essential part in the experiments, were they going to be the sort of people with whom I could work and live, and who could work and live together without constant bickering and all manner of (non-violent, of course!) quarrels? My assistant, Walter Bartley, was also a pacifist, and during the six weeks for which he had already been working he had shown himself to be an efficient worker and an otherwise normal person, but then he had worked in a scientific laboratory already, and also was being employed more or less in his own profession and not primarily as a pacifist guinea pig.

The first two men arrived on Saturday, the 25th of January, 1941. They seemed to be more or less normal people. One had trained and taught as a mathematics master in a secondary school, and the other had just completed a course at the Brighton Art School. They settled down quite quickly, and Mr. Roth, our Viennese chef, now had some more people to cater for. His first reaction to the volunteers did not seem too favourable—the first day I was informed that "that Peter" (the artist) was breaking all the china when he washed the dishes; two years later the culprit became Mr. Roth's son-in-law!

By the end of February 1941 our original full complement of 12 volunteers was in residence. They included all types of individuals and were drawn from a variety of professions. There was an electrician, a milk-roundsman, a shop assistant, a man

L. M. Rasmussen (ed.), *Human Guinea Pigs, by Kenneth Mellanby: A Reprint with Commentaries*, Philosophy and Medicine 134,
https://doi.org/10.1007/978-3-030-37697-0_6

who had tried a variety of trades from electric welding to winkle boiling, a ladies' hairdresser, a clerk and a baker. These were all full-time employees of the Institute; in addition, two other volunteers who held full-time jobs in Sheffield lived at the Institute and took as full a part in the experiments as their work would allow.

Our original plan was to offer volunteers conditions of employment which were economically on a par with those they would have found if they had been called up into the Army. Actually, they were always worse off financially than the worst paid soldier, though there were some compensations in their position. A basic rate of pay of 16/– per week was allowed, and in addition we contributed the whole of the cost of National Health Insurance and Unemployment Insurance stamps. "Board and lodging" was provided, but we did not supply any clothing or uniform. As a general rule we only intended to accept volunteers without dependants, so there was no proper scheme for marriage allowances.

The accommodation was provided for the volunteers in the Institute. The "drawing room," as it was originally, served as the common room; it was a very large room and took, without crowding, a long dining table, together with the settees and armchairs which we had purchased at the auction sales. At any rate, at the start, this furniture was for the most part in quite good condition, and the common room was much more comfortable than anything one finds in the ordinary Army billet. Sleeping accommodation was in dormitories, each with from three to five occupants. There was usually a shortage of drawer and cupboard space, which together with the natural inclinations of the volunteers generally meant that from the point of view of tidiness the bedrooms would have given any inspecting General apoplexy! Actually, when the scabies experiments demanded tidiness and rigorous segregation of each person's clothing and bedding there were no complaints as to the condition of the dormitories, so it cannot be said that the normal state of untidiness interfered with the work—or that it could not have been avoided if the volunteers had really tried!

The volunteers were employed primarily as subjects for experimentation, and it was understood that they must always be available when necessary for this purpose. It was intended that those who showed special aptitude should be kept busy in the laboratory or co-operating with the work in some way, and the others should do the house work and when not otherwise engaged should look after some allotments on which we intended to grow vegetables. I was, at the outset, worried by the possibility that there would not be enough work for the men to do and that "the devil would find work for idle hands". Actually, I was afraid that if we did not keep the volunteers reasonably busy they would get bored and miserable; I later found that some had immense powers of endurance when it came to doing nothing, but also there is no doubt at all that the house was happiest when some special job was on hand and everyone was rather overworked.

I did not know quite what I was letting my volunteers in for when I originally asked them to submit to infection with scabies. I knew that the disease could be very unpleasant and had heard patients saying that the intolerable itching had prevented them from sleeping for weeks, but it seemed unlikely that any permanent harm would be done to any of the volunteers. Nevertheless, at the outset I imagined that it would be a necessary safeguard for each man to sign some sort of detailed contract, setting forth his duties and the risks he was taking, in order to cover me in an emergency and to have something to which to hold a recalcitrant subject if he proved non-co-operative. When I made some investigations into the legal side of the question I gathered that certain risks were being taken by me in using human volunteers. No matter what he had signed, if any subject started getting nasty he could place us in an awkward position, but if our volunteers appeared to be co-operative and wrote and said they wished to take part in this sort of experiment we would probably be as well covered as by having all manner of complicated (and costly) legal documents signed and sealed and so forth. I know now that I have been lucky with my volunteers, but I have never felt that it would have been an advantage to have formalised our relations in this legalistic way. Indeed, I think much of the success of the experiment was due to its very air of informality. On the other hand, if work of this kind were to be carried out on a large scale and on a permanent basis I should agree, though somewhat reluctantly, that the legal side of the matter would have to be looked into very thoroughly, and full scales for compensation and disability would need to be worked out at the outset.

Frequently I have been asked how I "managed" the volunteers and made them carry out their various duties. I have always found this very difficult to answer. Those with an authoritarian or militaristic view of life have always assumed that in work of this kind Discipline with a capital D is necessary, and that you must never trust a man to do a job unless you see him do it yourself. This view is based very often on real practical experience with "volunteers" from the Army who have taken part in various tests and experiments. Of course the trouble here is in the nature of the volunteers and the reason for their taking part in an experiment. When a regular soldier volunteers to take part in an experiment he nearly always does so either because the sergeant-major says "I want three volunteers ... Jones, Smith and Robinson report, etc." or else because he thinks that by being a volunteer he will get a cushy job for a time, and perhaps some extra leave as well. He is not interested in the experiment itself or in the reason for carrying it out, and his lack of knowledge and interest make the most rigorous supervision necessary. Furthermore, once he has "volunteered" he is expected to continue with the experiment until he is dismissed. Finally, the military situation usually means that the Army is short of fit and intelligent men and only the crocks and the dullards are usually encouraged to volunteer for human experiments. But if they are treated as intelligent human beings, and are told what the experiment is about, I am sure that even soldier volunteers can usually be relied upon to give good co-operation.

As opposed to the military or disciplinarian standpoint, there are others who take a much more left-wing view of the matter, and who would consider that the subjects taking part as volunteers in a human experiment should be counted as equal partners, allowed a share in the management of the work and in the planning of the investigation. Although I am more inclined to agree with this view, I am afraid I feel that it is dangerous to take too democratic an attitude in a matter of this kind, and that it is essential to have one competent person in authority who is in a position to take all the important decisions regarding the planning of the experiments.

In practice, the organisation of the Institute was very much allowed to look after itself and was not based on any particular theories. It must be borne in mind that the volunteers had come entirely of their own free will—they had not been drafted or compelled to take part in the experiments. Had they not volunteered they would have been in other work which was allowed by their Tribunal's rulings, and would have at any rate received considerably more pay and probably (especially if they had accepted work in Civil Defence) even less work. Furthermore, there was no way in which we could hold a man to the job if he found the conditions intolerable, and so any volunteer could have left whenever he so wished. As it was, no one ever gave up before the originally suggested period had elapsed, and as a rule it was I who stopped any particular experiment, and not the participator, who was willing to continue as long as might be thought necessary.

One thing only was essential in the managing of the volunteers in these experiments, and that was to be willing to take the trouble to explain in detail just what was the purpose of anything which we asked them to do. I always made a practice of holding periodical meetings which they all attended, and where I gave a progress report and answered the questions which they had to ask. It is not always easy for the scientific research worker to explain to an audience which, though closely and personally concerned with the work, has essentially no scientific training or special knowledge of the investigations, but I believe that it is a useful experience and leads to clearer thinking on his part as well as theirs. Also the questions asked, though often not very profound, not infrequently proved useful in suggesting new lines for study. I think we usually managed to maintain a friendly atmosphere where the volunteers knew that they could come and discuss any matter concerning the research perfectly freely. As a rule, I took the volunteers entirely into my confidence about the progress of the work, though on occasion it was necessary to conceal some information, particularly when investigating such subjects as individual reactions to the presence of the causative mites and other topics with a considerable subjective element in them.

Although the work was always discussed with the volunteers, there was never any question regarding the final authority in planning what was to be done. In this I was completely autocratic. Later experience has shown that this is much the most satisfactory way to run work of this kind. It is best for all concerned for one person to be in the position to make every decision that is required; of course, he should take all advice and guidance that is available, but the thing to avoid is the delegation of authority to a committee, for when it comes to taking immediate decisions the joint intelligence of a committee seems often to be less than that of its least intelligent member! It is very valuable to have scientific committees to discuss scientific results and their practical application, and even future programmes, but they are not a suitable means for the actual running of an experiment; this is the one field where a Dictator is required!

Outside the scientific work, except that I insisted that each volunteer should be available when required for the work, I interfered as little as possible in the running of the household. The men themselves arranged the rota of household and other duties, and were responsible for all such work as the cleaning of the house. These duties were sometimes done well, sometimes badly, and sometimes apparently not carried out at all! I often felt the urge to interfere, and occasionally was compelled to do so, but on the whole the volunteers muddled through and even, at times, got things done surprisingly well. When I got exasperated during a particularly inefficient spell I consoled myself by thinking that we had recruited the volunteers for scientific purposes and not as housemaids or general servants, and it is true to say that whenever efficiency in the house was required for the experiments it was forthcoming. I suspect that if I had tried to be more "efficient" and insisted on "spit and polish" we might well have got more superficial appearance of cleanliness and efficiency but with a deterioration in the excellent co-operation in the more essential experimental work.

Not being a pacifist myself, and knowing very little about the philosophy and beliefs of pacifists other than what was contained in biased newspaper accounts of the proceedings of various Conscientious Objectors' Tribunals, which made the men appear to be impossibly opinionated, pig-headedly obstinate and incurably ignorant, I determined to avoid any discussion of the war, their beliefs or of pacifism with the volunteers. I had read a few of the writings of pacifist authors and these gave an even more unfavourable picture than that found in the anti-conchy newspapers, for the writers appeared, as a rule, humourless and intolerant. If the men were like these accounts written from either the favourable or the unfavourable aspect they were going to be pretty difficult to get along with.

The illusions contained in the last paragraph were soon shattered. Such views had been considerably shaken when I met several members of the committee of the Pacifist Service Units, for they seemed both normal and intelligent, but I was afraid that they might be complete exceptions to the general rule. It is true that many of the volunteers held views with which I did not agree, and when they got together in a bunch they sometimes produced what I thought were absurd pronouncements, but they were always redeemed by a reasonable amount of tolerance and a well-developed sense of humour. Their ability to hold their views with complete sincerity, and at the same time laugh at themselves, made life much easier for us all. Furthermore, none of them ever tried either to "save" me or to convert me to pacifism, which I like to assume was due to tact on their part, but which I realise may perhaps have been due to their realisation of my depravity and of the impossibility of such a praiseworthy task! I have, in fact, always avoided discussions about the war and similar problems, but mainly because I find these endless arguments of little interest.

The men's sense of humour was shown in a number of ways. It was felt that we needed a "coat of arms," and its design was discussed at length. Some of the volunteers thought it should include a yellow streak, to draw attention to their pacifist leanings. Eventually our artist produced a beautiful object, introducing for the first time *Sarcoptes* into the realms of heraldry and bearing the motto "Itch dien". The shield decorated the common room during the course of the experiment.

As soon as the first volunteers arrived they began the experiments which I had been planning for the past months. These first experiments will be described in the next chapter.

Chapter 5
How Is Scabies Transmitted?

The view which was most commonly held at the time when we started our experiments was that scabies was rife in the army, that when clean civilians got into uniform they rapidly became dirty and scabietic soldiers, and that whenever they went home on leave they took with them the disgusting parasites and so infected their unfortunate families! It will later be shown that this view could not be further from the truth, but it took us a considerable amount of time and trouble to elucidate that fact.

I now had at my disposal a dozen or so willing volunteers. They were all thoroughly examined by my colleague Dr. Andrew Wilson to ensure that they were fit and not suffering from any organic disease. Their skins were of course particularly scrutinised to make sure that we did not start with any volunteers already suffering from scabies. The time had come for them to be infected with the disease.

It has already been stated that much blame was put upon the Services for the prevalence of scabies, so that it seemed natural that the Services should be approached for infective material. I had made contact with the senior officers of the different Services in London, and Lt.-Colonel (later Brigadier) T. E. Osmond, who at that time combined the position of consultant on venereal diseases with that of consultant in dermatology at the War Office, arranged with the medical officers of the various army units in the vicinity of Sheffield that many useful facilities should be given me. In a later chapter the question of co-operation between the pacifist volunteers and the military is dealt with in more detail, so for the moment it is only necessary to say that we were able to obtain an ample supply of the bedding and underclothing immediately after it had been used by infected soldiers.

The plan of the first experiments was simple; we wanted to know just how long this infected bedding and clothing would retain its power to transmit the disease. This meant that it would be carefully stored under controlled conditions of atmo-

L. M. Rasmussen (ed.), *Human Guinea Pigs, by Kenneth Mellanby: A Reprint with Commentaries*, Philosophy and Medicine 134,
https://doi.org/10.1007/978-3-030-37697-0_7

spheric temperature and humidity, for it seemed likely that extreme cold in winter, or dry heat in summer, would kill any wandering parasites more rapidly than would moist equable conditions. We also wanted to know what was the minimum high temperature necessary for disinfestation, at which all stages of the *Sarcoptes* would be killed in a few minutes; much fuel was being wasted, and bedding damaged, by sterilization at unnecessarily high temperatures.

To start off with volunteers were put into beds made up with the blankets which these patients with scabies had used up till their treatment. The volunteers slept naked in between the blankets, which cannot have been an entirely pleasant experience, for though the army washed and sterilised all blankets at regular intervals many of the individuals who had just vacated the bedding were not particularly noteworthy for personal cleanliness. Other volunteers put on the thick army underclothing (vests, pants, shirts and in some cases khaki trousers) discarded by patients about to be treated, and these undergarments were worn constantly night and day for a period of at least a week.

I had assumed that as a result of the type of exposure described in the preceding paragraph, within a few days of using the bedding or clothing normal clinical scabies would be apparent in the majority of the volunteers. Therefore no one was more surprised than I when weeks and months, covered by a very large number of different experiments of this kind, went by, and not one of my volunteers had developed scabies. It seemed obvious that the current views on transmission were somehow at fault, and it was necessary entirely to change the whole plan of the experiments. The reactions of the volunteers at this time were interesting. At the outset they were all naturally a bit nervous about what was going to happen to them. They had volunteered to have scabies, and were all perfectly willing to go through with it but it was going to be an unpleasant experience. But now they had been doing the work of volunteers for what seemed a very long period, and no one had developed the disease or had anything unpleasant to report—they began to feel a bit fraudulent and all became exceedingly keen to develop and suffer from the disease. They had not, in fact, so very long to wait!

During this period, when we were daily expecting to find that all our volunteers were infected, and when none of them, in fact, was, we began to try and find out whether other methods than bedding were responsible for transmission. At first when a soldier was questioned he always said that he got "them" (scabies in common parlance is plural, there being presumably some hypothetic singular noun "a scabie") from the dirty blankets, and there would usually follow some long and tedious story about being on guard or in detention or somewhere where the bedding was used promiscuously. In civilian life doing duty as a fire-watcher, again a place where bedding was communal, was often blamed. Our main difficulty at this date was that we believed that scabies was a disease with only a very short "incubation period," and that the clinical symptoms which were apparent when a case came for treatment had developed as the result of infection in the last few days. As will be shown later, we eventually found that scabies often had an incubation period of up to two months, which makes it very difficult to trace the original source of infection. I think that it will be found that in most diseases the epidemiology of which is obscure one of the main troubles lies in a lack of precise knowledge regarding the period which elapses between infection and the development of recognisable symptoms; volunteers willing to suffer infection could obviously be of great use in elucidating many such problems.

There was one red herring which put us off the track for a few days. Major G. A. Hodgson had found that, in one large batch of patients admitted to a military hospital, those with scabies commonly possessed the lowest intelligence. Although Hodgson himself never suggested the idea, certain other people then stated that only dullards could be infected, and what was wrong with my experiments was not that the blankets, etc., were not infective, but that the subjects using them were too intelligent to pick up the disease. Further work showed that, although Hodgson's finding for a particular batch of soldiers was a true one, the reason for the result was concerned more with the extent of exposure of the different grades of intelligence than in the susceptibility of the individuals to infection. If a man picks up a mite he will develop scabies independent of his brain power.

Although the general view was that scabies was transmitted mainly through the medium of infected bedding, some workers had stressed the possible importance of close personal contact, and certain French dermatologists had somewhat characteristically classed scabies as a venereal disease (though as the highest incidence tends to occur in quite young children this would appear to suggest a somewhat jaundiced view of the morality of the young). So long as we believed that the disease developed in only a few days, it seemed very definite that at any rate some of the cases we saw in soldiers were not venereal in origin, for the men had evidently avoided any such contacts, either legitimately when on leave or otherwise, for several weeks. On the other hand many cases had a possible venereal origin, and quite frequently a patient reported that he had just come back from leave where his wife and children were under treatment for "itching spots". These cases looked like a proof of the short incubation period, and misled us for some time until we realised that during the earlier part of the war most soldiers got home for at least a week-end every six weeks, and scabies apparently contracted a week ago had probably been developing, undetected either by the man himself or his medical officer who inspected him, for many weeks before the last leave took place.

For a time the position appeared then to be that we saw many cases of scabies and we were beginning to doubt the current view that the disease had been caught from infected bedding. We seemed unable to infect any of our volunteers, which made it difficult to test out other types of transmission. I considered all manner of desperate possibilities. Would it be practicable to ask a pacifist volunteer and an infected soldier to share a bed, to see if transmission took place? And would it be possible to test out the venereal hypothesis? We discussed whether we should try to find some accommodating young woman with scabies to use as a source of infection. My volunteers became, I gather, a bit worried that they would be asked to commit adultery in the interests of science, and I was worried that if such an experiment were to take place the police might raid the Institute and I would be arrested for keeping a "disorderly house". Fortunately all our difficulties were solved at the crucial moment by the successful infection of two of the volunteers by less dramatic means. I suspect that some of the volunteers were a trifle disappointed.

Our first two cases of infection were due to the wearing of infected underwear. In both instances the garments were removed from patients who had been using them for a good many days, and when still warm from the body they were donned by the volunteers and kept on, as described above, for at least a week. It was well over a month before the infection was discovered, which was the first inkling we had that the view that scabies developed rapidly after the parasites had been picked up was not a true one.

Now that these volunteers were infected it was possible to get ahead with our experiments. We could not yet test the theory of venereal transmission, but it was possible to arrange for two individuals, one with scabies and one without, to share a bed. Allowing for the possibility of a long incubation period meant that fewer experiments could be carried out, but by the end of the summer we had performed 63 experiments with bedding and clothing and four in which infected and uninfected volunteers slept together. In 25 cases the use of bedding had given no infection. In 38 experiments with underclothing two volunteers had become infected. In the four bed-sharing experiments, three lasted for a week and gave rise to transmission, one lasted only two nights and gave a negative result. In these experiments the men wore pyjamas and there was much less intimate contact than on occasions where venereal transmission was likely to take place.

The conclusions of these experiments were clear. They suggested that bedding could not be the usual means of transmission, and that clothing, though more likely to spread the disease, could not be of any great importance because it is not a common practice for individuals to exchange their worn underwear whether or not they are infected with scabies. On the other hand, sharing a bed seemed to give rise to ideal conditions for transmission, and it seemed natural to assume that the more intimate the contact within the warmth of the bed the more likely transmission was to occur. Later work had supported these initial findings which have only required modification in matters of detail.

In our original estimate it will be seen that a sum of £20 is allowed for "Upkeep of horse" (see Chap. 2). The reason for this is that we wished to carry out a series of laboratory tests of the viability of the *Sarcoptes* away from man; thus if we found that under the most favourable conditions of temperature and humidity the creatures never lived longer than say 10 days, then if infected clothing and bedding could be stored unused for that period they would be perfectly safe to use again. These experiments were of course planned at the time when we thought that bedding was the main cause of the spread of scabies. Now for laboratory experiments of this kind very large numbers of mites are required, and it appeared unlikely that sufficient could be obtained from human subjects. When a horse, however, is infected with its own particular *Sarcoptes* mites can be removed from the crusts on the skin in thousands. It was proposed to carry out extensive experiments on these horse mites, and to determine accurately how they survived under a variety of conditions. Finally these results would be checked by using the genuine human parasites; such a check would require far fewer mites.

But, as Mrs. Beeton might have put it, "First catch your horse." One type of mange in horses is caused by this *Sarcoptes*; the disease is then known as "sarcoptic mange". There are, however, several types of mange in the horse, usually caused by mites only distantly related to the species in which we were interested, and unsuitable for our experiments. I could not discover a horse with sarcoptic mange anywhere in the whole of Britain, and I doubt whether one exists today. The officials of the Ministry of Agriculture tried, and so did the Army Veterinary Corps, which in the last war was greatly troubled by this disease, but with no result. We were often on the track of such an animal, but in the end it had just been slaughtered, or had the wrong disease—or it simply vanished into thin air. Eventually it became clear that we were going to have no horse to maintain. Fortunately the rest of the work had developed in such a way as to make the horse not completely necessary. First our experiments with bedding and clothing had shown that even under the circumstances most favourable these articles seldom transmitted scabies, so it was only of minor importance to know how long the mites could survive in conditions they seldom occupied. Furthermore, though it was a very tedious procedure, we could obtain sufficient human mites to carry out the majority of the originally planned experiments.

These experiments on the biology of the *Sarcoptes* away from its host, man, showed that it was not a very hardy creature. A temperature of 120 °F (49 °C) killed all stages in ten minutes; such a temperature can easily be reached in any disinfesting apparatus, and will not damage fabrics. Under the conditions of a warm airing cupboard all mites died in under two days. When kept cool and damp, which favoured survival to the greatest extent, one single parasite out of several hundred lived for a fortnight, but most died in a few days. It appears that mites seldom find their way on to bedding and clothing, and it is obvious that even if they do so they will not remain alive and keep it infective for very long.

A few other experiments on mites infecting animals were carried out. Dr. C. M. Scott, head of the pharmacological laboratory of I.C.I. at Manchester, who supplied us with a number of substances used in treatment, also very kindly sent a hamster (a small mammal about the size of a rat) infested with the mite *Notoedres*, a parasite somewhat resembling *Sarcoptes*. Although the mites from the hamster were different from those on man, handling them soon taught us the technique required for dealing with creatures of that size (i.e. a sixtieth or a hundredth of an inch in length), and this proved invaluable when we came to tackle *Sarcoptes*. We also collected different mites from rabbits, cats and other animals.

Our early experiments on transmission produced scabies in what may be described as a "natural" way, for the mites causing infection were picked up either by personal contact or by using clothing containing the parasites. Later it proved more convenient to use a process more or less of inoculation. We had learned by this time to find and remove the parasites from patients as required, and these could be placed on the skin of patients and allowed to burrow in whatever position we desired. These infections could then be kept under close observation, and all signs and symptoms related to the original time and intensity of infection. Incidentally, these infections developed into clinical scabies quite indistinguishable from the disease as contracted in the "normal" manner, whether in volunteers or in patients coming to seek treatment.

The experiments recorded in this chapter were the main investigations as originally planned, and they were for the most part completed before the initial grant was exhausted. Various circumstances, such as the blitz, which delayed our start without stopping overhead expenses, caused us to spend the original sum rather sooner than anticipated, and so we were in need of a further grant after ten and a half months of experiments. The Ministry of Health again agreed to finance the research and made us a further grant which was to continue the work until the end of March 1942.

Chapter 6
Dietetic Experiments

The investigations soon settled down to a matter of routine. Every day, Sunday included, every volunteer's skin was given a thorough inspection. This meant that he stripped completely and his whole surface was carefully scrutinised, particular attention being paid to any spots or blemishes; these were examined with a hand lens or a binocular microscope. Every lesion which could be attributed to scabies was noted, and for much of the time the presence of the actual parasites themselves was determined. I soon came to know thoroughly every inch of each of the subject's skin, and could have recognised any of them by the physical or anatomical peculiarities of any part of the body, whether usually visible or invisible.

These frequent and thorough inspections took a good deal of time and they gave a considerable amount of work, but they did not absorb all the energies of the volunteers. As a rule, not all the subjects were infected with scabies and even when infected they were otherwise in good health. I decided that we were not making the maximum use of our opportunities. Here we had a group of intelligent and willing human volunteers, living under closely controlled conditions ideal for experimental work on a variety of subjects. I discussed the possibility of some dietetic experiments with the men, and they were quite enthusiastic about the possibility of combining some such work with the scabies investigation.

Next I discussed this possibility with Dr. H. A. Krebs, who is in charge of the department of Biochemistry in the University of Sheffield. He had been concerned with a good deal of dietetic research and I hoped that he would be prepared to take a considerable part in the direction of the sort of experiments we were considering. There were very many problems which we could possibly tackle, but we felt that the best possible use must be made of the volunteers and that we should only consider working on something of real importance. Our volunteers being pacifists, we could

L. M. Rasmussen (ed.), *Human Guinea Pigs, by Kenneth Mellanby: A Reprint with Commentaries*, Philosophy and Medicine 134,
https://doi.org/10.1007/978-3-030-37697-0_8

obviously not work on any question purely of military importance such as the production of emergency rations for mobile troops, but there were many questions regarding the nutrition of the civil population either of Britain or of occupied Europe which required elucidation. It should be remembered that at this time, early in 1941, shipping losses were high and the future food position for Britain was not thought to be particularly favourable, and widespread malnutrition was foreseen unless the food situation was carefully watched.

An official "Food Policy Committee" had been appointed, under the chairmanship of Sir William Bragg, President of the Royal Society (who incidentally knew all about the volunteers, as he was chairman of the Sorby Committee). I wrote to Sir William explaining that we had a dozen volunteers eager and willing to carry out dietetic research, and that we thought the conditions were ideal for such an experiment. I had of course made sure that the Ministry of Health would not object to the volunteers being used for dietetic work, making it quite clear that we would consider the scabies work of primary importance and not allow it to be interfered with in any way by other investigations.

The Food Policy Committee considered our proposals and added that "we greatly admire the spirit of your volunteers." They did not themselves put forward any specific problem, but handed over the suggestion to the Medical Research Council. For certain personal reasons I had hesitated to approach them in the first instance, but it was obviously much more satisfactory to be in close touch with the organisation which is responsible for carrying out so much of the medical research in this country.

In a few days I heard from the secretary. He mentioned that Dr. R. A. McCance and Miss Widdowson, in Cambridge, were doing some important nutritional experiments on human subjects and suggested that we should collaborate with them. We were also warned that human dietetic research was a very difficult and tricky field of investigation and should not be entered without learning the technique and the difficulties; we would learn all this best by discussing the whole question in detail with the Cambridge workers, and by studying the whole lay-out of their experiment.

Soon after this Dr. Krebs and I paid a visit to Cambridge. We learnt that Dr. McCance and Miss Widdowson had been studying calcium balance and other problems in a group of volunteers for many months. The volunteers had been living on a restricted diet, prepared in the laboratory by a qualified dietician, and for long periods everything eaten and drunk had been weighed and samples submitted to detailed analysis. All waste products of the volunteers had also been collected and analysed, so that it was possible to produce a complete "balance sheet" for each person with an exact knowledge of just how much of each substance had been taken in and how much put out by the body. The results of McCance and Widdowson's experiments were of considerable importance, and also revealed that there was an extensive field for further investigations on allied topics.

One thing is completely necessary for dietetic experiments of this kind, and that is that the volunteers should be absolutely reliable. It might be possible to treat human beings in a gaol or a hospital like experimental animals, and supervise rigidly what they eat, but they would not be living under anything approaching normal conditions. Those who had not had personal experience of our volunteers have at times wondered if they were sufficiently closely supervised, but I feel that the best controller in work of this kind is the man's conscience—if he has willingly volunteered to go in for an experiment, and is properly treated, he will not cheat. We could not supervise our volunteers all the time and had to rely on their word that they had adhered closely and accurately to the diet and to the quantities recorded. (In addition, our analyses of excreta would soon have made us suspect any unreliability on the part of the subjects.) But in the dietetic experiments, as with the scabies work, the volunteers were closely interested in the scientific work, and if any had been inclined to stray he would have felt that by doing so he was not only endangering our investigations, but also spoiling his own work and that of his fellow volunteers. Provided that the volunteers look upon the experiments as their experiments, and they are not treated as mindless guinea pigs, it is not difficult to get complete co-operation. It is further essential to avoid at all costs any suspicion in the volunteer's mind either that the research workers in charge of an experiment are getting bored with it, or that they are incompetent at their job and are asking people to put up with unpleasant experiments for no practical or scientific purpose.

After considerable discussion with the workers in Cambridge, we decided that it would be possible to do some work on similar lines in Sheffield. Dr. Krebs was responsible for the scientific direction of this investigation, while advice was received on many practical points from Dr. McCance, Miss Widdowson, and others. Before this I had been the only scientific worker who had been in close contact with the volunteers, but now Dr. Krebs had to cope with them also. After a short time he came to enjoy their complete confidence, and everyone appreciated the tactful way with which he handled them and their problems, but at the outset, probably due to a lack of explanation on my part, he ran into a number of unexpected difficulties.

Many people, scientists as well as others, appear to have a good deal of difficulty in understanding the point of view of the pacifist. This question is dealt with more fully in a later chapter, but for the moment one or two special aspects may be mentioned. The volunteers all realised that in peace-time it is unlikely that funds would have been available for this type of work and that it was therefore being supported partly due to its relation to the war effort. They were willing to co-operate, however, because the work was mainly directed to scientific and humanitarian ends, and would have been worth while carrying out if the war had not occurred. When mentioning new experiments it was necessary to stress this humanitarian outlook, even though they knew, and knew that I knew they knew, that in some ways all research was directed to military ends, or at any rate it might give results of military importance. On the first occasion that Dr. Krebs talked to the men he did not adhere to these conventions, and explained how important work on wholemeal bread might be in the saving of "shipping space," presumably to allow the importation of more armaments for the efficient prosecution of the war. This caused a temporary upheaval in the household, but when it could be shown quite truthfully that in the long run this sort of work was worthwhile, whether or not it helped the military, then there were no further difficulties about proceeding. I do not think that this all means that the volunteers could be charged with hypocrisy (though some of the men themselves have been worried by this point) and at later dates there have been other proposals for more purely military research with which they might well have refused to co-operate, but which I myself did not feel justified in asking them to carry out. It is essential not only to explain scientific details to this type of volunteer, but also to respect and be tolerant of his beliefs. I even found myself developing a "conscience" regarding the sort of experiments we should, or should not, carry out.

For the first dietetic experiments we found that it was only possible to cope with six volunteers—the numerous analyses performed made a large number impracticable. A certain amount of preliminary training of the volunteers was needed, after which the investigation continued for the greater part of the year. It did not involve anything at all heroic, or even uncomfortable, but it became a very tedious performance.

These experiments dealt with the properties of bread, and in particular of the "National Wheatmeal" flour. At the date of this experiment there was no restriction as to the type of bread sold in the shops, white bread being abundant and normally obtainable, but the Ministry of Food had just introduced the national wheatmeal loaf which later entirely replaced white bread. When wheat is turned into flour, only a part of the whole grain is used, as a rule. Strictly speaking, "wholemeal" bread should contain 100% of the wheat, but most so-called wholemeal contains a slightly lower percentage. White flour contains a much smaller percentage of the grain, in the region of 70%; the rest of the wheat is used for feeding animals. Now, at first sight, it would appear economical to use true wholemeal, where each hundred pounds of grain yields a hundred pounds of flour. However, man cannot digest every part of the wheat, and it has been said that it is more economical to use a "low extraction" (i.e., well under a hundred per cent of the wheat) flour and feed the residue to animals which can utilise it. For national economy it was important to make the maximum possible use of the wheat available, and make a flour of the highest possible extraction compatible with digestibility.

In addition to this question of digestibility, the nature of the bread may have an influence on the calcium metabolism of the population. The mineral calcium is essential among other things for the proper formation of bones and teeth; insufficient calcium in the diet means improperly formed bones and is very dangerous in childhood or pregnancy. The question of calcium metabolism is a complicated one, and much work has been done by a host of scientific workers on the relation between calcium and other minerals in the body, and on the action of the various vitamins, but one salient point is that not only must the diet be sufficiently rich in calcium, but also the calcium must be absorbed by the body. A normal adult should be in "calcium balance," which means that his diet should contain the same amount as is lost by excretion. If he is in "negative balance" and excreting more than he takes in, it means that the reserves, particularly in his bones, are being depleted (it is not usually realised that bones are living structures and diet and other factors can make them denser or lighter even in adult life). A negative calcium balance will eventually lead to disaster. An adult in "positive balance" is adding to his reserves; growing children and pregnant women must be in positive balance, though a woman will give up large quantities of her reserve calcium for the benefit of her child. Now cereals have been shown to affect calcium balance, and the more cereals in the diet on the whole the less calcium tends to be absorbed. Recently, this effect of cereals has been shown to be connected with a substance called phytic acid, which forms an insoluble compound with calcium, so that the more phytic acid there is in a diet the less of the calcium the body will be able to absorb. McCance and Widdowson's experiments had dealt with this point and they had shown that a high extraction (i.e., "wholemeal") bread, containing a lot of phytic acid could, in human beings, easily produce a negative calcium balance. We wished to find out whether the national wheatmeal, which contained 85% of the wheat and therefore represented a compromise between wholemeal and white bread, had much effect on calcium balance.

To study the effects of wheatmeal bread on their metabolism, the volunteers were given a greatly restricted diet. Substances rich in calcium, like milk and cheese and certain vegetables, were greatly curtailed or entirely excluded. It was aimed at making half the energy value of the food taken come from bread, which meant eating considerably more than a pound daily. To make the necessary comparisons periods were spent on white bread. One difficulty was that it was usually necessary to estimate how much bread was to be eaten in any day at breakfast time, and where a volunteer's "eyes were bigger than his stomach," it might mean a struggle to get the remaining slices down at supper time.

Everything eaten was weighed. This was easy for bread, butter and jam, which were weighed out in advance, but not so easy for the food consumed at dinner. In addition to weighing food, a portion equivalent to an exact fifth was always retained for analysis. This meant at dinner time two accurate weighings of meat, of potatoes, of any other vegetable and gravy. These eight estimations might easily take a quarter of an hour even after practice, so a properly hot meal soon became a fond memory. Drinks were taken in graduated beakers, so that their measurement was comparatively simple. The samples of the foodstuffs were specially treated and subjected to analysis in the laboratory.

At the same time as the food was weighed and analysed, all the excrement had to be collected. While they were about the house this was not difficult, though as it had to be weighed and measured the job was not a pleasant one. However, if one of the volunteers went out for the day the trip became a major operation. All food had to be weighed, sampled for analysis and packed up, for nothing could be purchased outside. It was permitted to drink outside, provided that the drink did not contain milk and that a sample of the liquid was measured and retained for analysis. Then further jars had to be carried for the urine and faeces, and occasionally one had the spectacle of a volunteer cycling hectically home with full containers and much discomfort as he had underestimated his fluid production.

The results of these investigations were quite interesting. It appeared that the National Wheatmeal flour approached white flour in its digestibility, and that in its production almost the maximum amount of the wheat that man could digest had been retained with very little added indigestible material. Furthermore, the amount of the phytic acid (the substance which tends to render calcium unassimilable) was less than might have been expected, far less than in wholemeal, and the composition of the flour was such that a further proportion of the phytic acid was destroyed when the bread was baked. We were able to suggest what amount of calcium should be added to the bread if it were felt necessary to neutralise this phytic acid in this way.

This experiment gave some quite useful results, and also proved of use in training some of the volunteers in the technique of dietetic work, a training which proved invaluable in later work. The scientific direction was all in the hands of Dr. Krebs, and I think that the experience was useful to him in showing him the viewpoint of a section of the population which he had not previously studied; he soon became very adept at getting the fullest co-operation from the men, so that the experiment proceeded smoothly and efficiently. At the same time the scabies work had been progressing quite unaffected by the dietetic experiments.

Chapter 7
Co-operation with the Military Authorities

The general question of the way in which various sections of the community react to pacifists is discussed later, but in this chapter it is proposed to deal with the way in which this experimental work, so largely carried out by conscientious objectors, was related to the work of the military authorities.

At the outset arrangements were made to obtain "infected blankets" from the beds of soldiers infected with scabies. Local medical officers were most helpful, in particular Captain Lougee, who was in medical charge of a local training unit which had a large intake of new recruits. At this stage the military authorities in Sheffield did not treat their own cases of scabies; they either sent them away to hospital, or to the cleansing centre run by the City Health Department. As mentioned above, after the "blitz" we provided temporary premises for this cleansing centre and also did the actual work of treating male patients. A certain number of soldiers was included in those treated, and when the city authorities opened their new centre and transferred their staff we arranged to continue with the treatment of Service patients.

At first, these soldiers with scabies were just treated during a period of an hour which they spent at the Institute. They arrived with their bedding, and I had been allowed to draw a considerable quantity of blankets and underwear from the local army store to replace articles left with us for experimental purposes. We supplied each man with clean blankets and underclothes after his treatment. Those which he had been using were employed in the experiments described in Chap. 4. Not infrequently one saw a pacifist volunteer, who had refused to wear uniform or join the army, clad almost entirely in khaki garments and wearing them with a good grace.

Soon we felt that it would be advantageous to have the cases of scabies more under our control so that we could ensure proper treatment and also more uniform "infection" of their possessions. We therefore turned one of the bedrooms into a military ward. Having this military ward in the house we had the interesting situation

L. M. Rasmussen (ed.), *Human Guinea Pigs, by Kenneth Mellanby: A Reprint with Commentaries*, Philosophy and Medicine 134,
https://doi.org/10.1007/978-3-030-37697-0_9

of pacifists and soldiers living in amity under one roof, with the pacifists waiting on the needs of their military guests. Feeding the patients was sometimes difficult due to rationing, but we arranged to draw the men's rations from the appropriate store. Mr. Roth, our Viennese chef, nearly always complained about the quality and quantity of the food supplied, and it was certainly a bit erratic, sometimes comprising enormous quantities of bread and little else, while at other times large amounts of almost-forgotten luxuries made their appearance; on balance, I think that the household received a good deal more than it supplied in the way of food.

The soldier patients showed no objection to being admitted to the Institute; in fact, they seemed to enjoy their stay! They did at least get a proper bed to sleep in, with real sheets; this was greatly appreciated, and for the first 24 hours most of them were sound asleep when they were not being treated. Treatment was on the same lines as those used by the Health Department—a good long soak in a really hot bath with plenty of soft soap and vigorous scrubbing of the skin, followed by a liberal plastering with sulphur ointment. This procedure was repeated on three successive days, and between treatments the patient remained in bed. One of the volunteers was responsible for the majority of treatments and he became very adept at it, and also at catching the patients when they fainted, which many "tough" soldiers did when they stood up after the bath. This treatment proved very effective, and during several months there was not a single relapse (later work showed that this was not surprising).

Although it proved so satisfactory to admit soldier patients with scabies to the Institute and to treat them there, we were limited for space and other accommodation, and could only take a maximum of eight cases a week. This, at any rate at first, gave us enough "infected" bedding and clothing for the experiments on transmission, but insufficient patients for any proper comparison of the different methods of treatment. We therefore tried to find a method of obtaining larger numbers of patients for treatment and for careful scrutiny over longer periods than were at present practicable.

Ideally we required an unlimited number of cases of scabies who could be treated in any way we liked and who could be kept under observation more or less indefinitely. Curiously enough, this was very much what we got in the end, though when I started out I never dared to hope that such facilities would be forthcoming.

The military authorities throughout the country were getting very worried about scabies, for the sickness returns for the first year of the war showed that it had caused a shockingly large amount of absence from duty. Soldiers were, like civilians, being admitted to certain hospitals for treatment and being detained for many weeks. Furthermore, some civilian public health departments were putting the blame on the army for spreading scabies and causing epidemics among their civilian populations. Under these circumstances they were extremely willing to co-operate in any work which might possibly relieve the situation.

In the early summer of 1941 I had an interview with Major-General David Richardson, Director of Hygiene in the Army Medical Service (it should perhaps be explained that although army medical officers belong to the Royal Army Medical Corps, when they reach a rank higher than a lieutenant-colonel they go on to the staff and become officially "late R.A.M.C."). I explained to him what we were doing, how our smallish number of patients was insufficient for any proper assessment of any type of treatment, and put forward my hypothetical and ideal plan for solving these difficulties. I said that what we would like would be, in effect, a small hospital with up to 30 beds solely for the treatment of scabies; somewhat to my surprise, General Richardson gave his approval to this idea and promised to do all he could to make it a reality.

Sheffield was situated, for military purposes, within the territory of the Northern Command, and I had several interviews with people at Northern Command Headquarters in York about our proposed scabies hospital. In order to hurry matters up I went snooping round myself and found a large and convenient house, only a few minutes from our Institute, and used at present for billeting troops. The house was in good structural repair, though the woodwork and paint had suffered in the usual manner from those who had been billeted there. The local unit whose troops were billeted did not really want to give up the house, but at the time they had more requisitioned buildings than they could use and this difficulty was easily overcome. The place was painted and decorated and fitted up with beds and all necessary furniture and equipment ready for 20 patients and appropriate staff in a matter of six weeks. Soon afterwards it was equipped for a further ten patients. For purposes of administration the place was considered to be part of a Camp Reception Station (which is, in effect, a small military hospital where minor ailments are treated and patients may be detained for a limited period). The medical officers and staff of the Camp Reception Station were generally very helpful, giving all assistance over administrative matters, but not interfering at all with our research.

Our scabies hospital was given a staff of R.A.M.C. medical orderlies, a corporal and several privates, and these men were taught by us to assist with the investigations; several of them gave a great deal of most valuable help. But much of the work in this military establishment was carried out by pacifists, by my assistant Walter Bartley, and by several of the volunteers, who soon became very familiar with scabies and the *Sarcoptes* which caused the disease. These soldiers who had been sent to staff the hospital always seemed to get on very well with the pacifist volunteers, and the pacifists carried out their "military" duties in an exemplary manner. Not infrequently, when off duty, the army medical orderlies were to be found in the Institute having tea with the pacifists or taking part in some other social activity.

Official instructions were sent to all army units within a wide radius telling them to send all their cases of scabies to us in Sheffield for treatment. They soon began to arrive, and at the peak of our activity we saw as many as 150 patients in one week. On arrival each patient was thoroughly and carefully examined to see whether he had, in fact, got scabies; nearly 25% were not so infected. We did not pretend to be able to diagnose or treat other skin diseases than scabies, and only set out in this preliminary examination to find or exclude infections with *Sarcoptes*. Difficult cases which were apparently not scabies were sent to the Dermatology Department at the Sheffield Royal Infirmary for examination by the physician in charge, Dr. Rupert Hallam. Dr. Hallam was always very helpful about these men, and throughout the whole investigation gave us the benefit of his very wide practical experience on many occasions.

When scabies had been positively diagnosed the patient was given his treatment. This was the most important part of the investigation from the point of view of the army, and to establish confidence in the various units from which the men were drawn it was essential at first that the treatment should always, or nearly always, be successful. So for several months we continued to use sulphur ointment, giving three baths and three applications of the medicament as described earlier in this chapter. As before, the results were almost invariably good.

I am afraid that I do not know just what the patients admitted to our scabies hospital thought about the place. Probably a lot of them just did not think at all. As a rule, when a man is in the army he seems to accept everything that happens in a fatalistic way, and as he feels that he has no control over his movements or his destiny he takes little thought and less action to attempt to improve or alter his circumstances. The men were grateful for their treatment and enjoyed a comfortable rest from more strenuous military duty.

After our routine had settled down at the military treatment centre we became a little more ambitious and tested out a whole series of remedies and treatments. We were becoming more familiar with the itch mite, and learned to observe these creatures in all our cases and assess the effect of the different treatments, not only on the clinical manifestations of scabies, but on the actual causative organism itself. During some of our tests, where patients had every square inch of their skin examined under a magnifying glass, they must have often wondered what it was all about. One of the volunteers also put most of the patients through an intelligence test, as we were trying to find if any particular type of individual was especially liable to the disease.

An ordinary case remained with us for a period of three days, which, at the outset, was considered a very short time in which to cure scabies. It was necessary at times, however, to keep cases which had had special forms of treatment for much longer periods, up to several weeks. This might have proved unpopular with the commanding officers of the various units, but to avoid any such difficulty we took great care in our selection of patients for detention for longer than the usual period. N.C.O.s were always discharged with the greatest possible speed, as were others who occupied key positions in their units. This meant that we could not use the best type of men for these tests, and it was obvious that if they had been employed as "volunteers" for any medical experiments constant supervision would have been quite essential.

Our treatment centre was entirely staffed with males, and the first time any female patients from the A.T.S. were sent for treatment it caused both amusement and embarrassment. Diagnosis was, fortunately, easy and could be made without any intimate examination, but it was thought impossible to have treatment given by our orderlies, and anyway such treatment is forbidden by army regulations. Fortunately it was always possible to obtain the services of an intelligent V.A.D. and to treat the girls as out-patients.

The few A.T.S. patients drew our attention to the problem of scabies in females, where it was causing almost as much trouble as were head-lice or "nits in A.T.S. nuts." Our facilities were not really convenient for the treatment of these girls, and we could only just treat them without being able to obtain any useful scientific information about the disease in women, where it might, we thought, present some special problems. I therefore put forward to the War Office a scheme for a special treatment centre on the same lines as the existing place, but for female patients. This proposal also was agreed to and, after a longish delay, we obtained premises where 12 girls could be treated as resident patients; this centre was staffed with V.A.D.s and A.T.S. orderlies. It was possible to make a thorough study of scabies in the female and to obtain other useful information about the head-louse.

In addition to all these invaluable facilities, the army also helped in the question of scientific staff. In September 1941 Major C. G. Johnson, R.A.M.C., was seconded to work with us in Sheffield. Major Johnson is an entomologist whom I had known for a good many years, and we had both worked in the Entomology Department at the London School of Hygiene and Tropical Medicine. He had previous experience of parasitic insects and soon became familiar with scabies, and added considerably to our knowledge during his period of eight months in Sheffield.

It was most convenient to accommodate Major Johnson in the Institute, so that we now had an army officer in addition to the pacifist residents. I at first thought that some of the volunteers might not like such close contact with an officer, but no difficulties of any kind arose, for he was not in any way an aggressively military individual and he always handled the volunteers in a most tactful fashion. We had a lot of military visitors at the Institute, and they were always interested to meet volunteers and to learn at first hand their views, based on such intimate knowledge of scabies! Even regular generals emerged quite favourably from the critical scrutiny of the pacifists, who soon found that in their present position they were meeting far more senior officers than they would have done had they actually joined the army themselves.

A good many scientific workers have been very critical of the military authorities, who have been stated to be unsympathetic to research and unlikely to accept at all readily any new discoveries. In the scabies investigation I certainly did not find these criticisms to be at all justified. I have already described the remarkable readiness with which the army authorities provided facilities for our investigations, and when we had some conclusions these were put into effect. For this work I found that being a civilian was a considerable advantage, for it meant that I was in a position to approach directly the most senior officers without in any way offending military tradition or discipline. I have since found that certain senior officers who were most co-operative to me as a civilian are by no means so helpful to a scientific worker in the uniform of an officer of lower rank.

Even in the scabies investigation there were, however, occasions on which the military mind was manifest. Major Johnson was expected to send in progress reports to his head-quarters at fairly frequent intervals; this was quite a good idea, as it helped to keep us informed exactly how the work was getting along. On one occasion, however, he was rash enough to state that we had observed a certain proportion of crab lice among our scabies patients. (The crab louse is a blood-sucking insect parasite of man usually found among the hairs of the pubic region, and most frequently spread by sexual intercourse.) He thereupon received an official rebuke, pointing out that he was supposed to be investigating scabies only and would he therefore devote his entire attention to that subject? As the work was being carried out in my laboratory I felt compelled to protest and ask whether this meant that, although he was intended to study the *Sarcoptes*, which could commonly be found burrowing in the skin of the genitals, he was nevertheless expected to turn a blind eye upon any other parasites a few millimetres away among the pubic hairs? There were no further difficulties on this score, but I cannot imagine any civilian in a position of authority causing trouble on a matter of this nature.

As well as obtaining so much help from the military authorities, we tried to give them some direct assistance with their scabies problem, in addition to communicating the results of our various experiments. On a number of occasions groups of medical officers and other medical personnel came to the Institute for courses on the diagnosis and treatment of scabies. The first of these courses took place in November 1941, when five medical officers, one from each army command, attended in Sheffield for one week. The theory was that these officers should learn all that we could teach them and then return to their respective territories to spread the gospel.

The officers selected proved to be apt pupils and I think that they appreciated the course and found it useful. When they first arrived, however, I got the impression that they had been sent to Sheffield without having themselves been consulted and without much prior interest in scabies, and, in fact, they rather wondered "what the hell it was all about anyhow." Although we were only faced with a small class of five, I decided that it would be best to run the course quite formally, but with plenty of time for informal discussion. We also wanted to give a good deal of practical instruction, for which reason a larger number of medical officers could not very well have been accommodated.

Each day the course started with a lecture on some aspect of scabies as affected by our research, after which the officers proceeded to the army treatment centre for instruction in finding and removing parasites from patients suffering from the disease. This completed the work for the morning. I considered that this practical work with actual soldier patients was of primary importance, and that if we could do no more than teach these officers to be able to check their diagnosis of every case of scabies by finding and removing the causative itch mite our time would have been well spent. Actually the officers soon learnt this trick, at which none of them had been previously adept, and they then became much more keen on the whole subject and found it much easier to accept what we were trying to teach them.

The majority of lectures were given by Major Johnson and myself and we also supervised the practical work. Much of the actual practical instruction was, however, given by our pacifist volunteers, so that we had the interesting and unusual spectacle of conscientious objectors instructing army officers—it did them both quite a lot of good. For these courses we were fortunately also able to obtain some excellent help from outside. One visit was paid to the Skin Department at the Royal Infirmary, where Dr. Hallam gave a talk on the treatment of secondary infections, and Dr. L. E. Robinson, the chief biologist of the Cooper Technical Bureau and a well-known authority on acarine parasites, came and gave a lecture on "Scabies in animals". For this lecture we invited the officers of a nearby Army Veterinary Corps unit, and we were overcome with admiration for their uniforms, for the veterinary officer in his riding boots, immaculate breeches and gleaming metal looks so much more like a real soldier than the present-day utility model!

I like to think that it was entirely our technical instruction which made these courses "go," but I think that things were helped a little by our attempt to combine work with entertainment, and we made some slight effort to look after the social life of our pupils outside working hours. In the army men get so used to being sent on courses that occupy them during the day but make no provision at all for their entertainment, which means that they are just dumped in a depressing billet in a town with apparently no facilities for amusement, that they generally appreciate any attempt to entertain them. Lest I should be thought to be libelling Sheffield, I hasten to add that the Lord Mayor stated recently that it was "not such a dull place" and that there were many "voluntary bodies whose object was the entertainment of the troops". I suspect he didn't mean quite what we read into his statement, though either way it is no doubt correct. Unfortunately, some of the "voluntary bodies" tend to play their part in the spread of scabies!

Although the British Army sent a number of officers for these courses, and for shorter periods of instruction, it was the Canadian medical authorities with which we had the closest co-operation. I have always been greatly impressed with the efficiency and keenness of the Royal Canadian Army Medical Corps. Their officers not only appear to be interested in their jobs, but also do everything they can to keep fully up to date, and I have always had the impression that they make the best of the military system (which, when all is said and done, has its advantages) without being smothered with red tape or overcome with the intellectual barrenness so common in other armies. I believe that there are no regular soldiers in the Canadian army medical service.

We ran several courses for the Canadians. They would send a number of men ranging in rank from lieutenant-colonel to private, and they sent people who were not only really interested in the subject of scabies but also in a position to put what they learned into effect as soon as they got back to their units. As a result, certain hypotheses which we put forward were tried out on a large scale among the Canadian forces, so that we obtained from them a lot of most useful information at a time when it would have been otherwise impossible to procure. I also had the opportunity of visiting a good many of these Canadians who had been to Sheffield in their own units, where I not only saw that they had carried out our ideas most efficiently, but also enjoyed their most lavish Canadian hospitality.

When United States troops started coming over to the British Isles we soon made contact with them. Much earlier we had been visited by the staff of the Harvard Hospital, which came over to Britain in the early days of the war, when the U.S.A. was still neutral, and many of whose medical officers later joined the American forces. At the invitation of American Army Headquarters I visited Northern Ireland and lectured to the medical officers there on scabies—at the time it was said that there was little scabies in the U.S.A. or in their troops when going overseas, but it was feared that when they reached Britain they might develop a high incidence. They did not, in fact, develop much scabies, and it is possible that this investigation helped a little in preventing more infections from taking place.

Chapter 8
The Work Goes On

The routine of the scabies investigation continued uninterrupted by dietetic experiments or by the visits of army officers. The membership of the household changed a little from time to time, though after three-and-a-half years some of the original volunteers were still in residence and taking part in experiments. After the first nine months we decided that it would be possible for the volunteers themselves to be responsible for their own catering, and Mr. Roth departed to take charge of a British Restaurant in a nearby town. The ordering and preparation of the food was in the charge of two of the men, one of whom had experience of the grocery trade while the other was a professional baker. With dietetic experiments going on it was more satisfactory for us to have full control of the food, for though Mr. Roth was a highly skilled cook I was not sure that he fully understood the importance of the dietetic experiments and was afraid that he might forget and add some forbidden substance to a dish in order to make it more palatable.

When the men took over the cooking the results were a little erratic. On occasion the food was excellent but at other times it was not quite so satisfactory. The unfortunate thing was that the best cooks were as a whole the same individuals who were efficient at laboratory work and at any other jobs. I have always found that as a whole providence has distributed talents to people in a most uneven way, and that the man who is good at one job is also reasonably efficient at the majority of other tasks. This was certainly true of the general run of our volunteers, and at times it meant that if we wished for full efficiency in the laboratory or treatment room we had to be prepared to submit to a dietetic régime which left much to be desired. But in the end most of the men became, if not skilled chefs, at least able to produce a meal that their comrades were able to consume. Most of the volunteers put on weight during their stay in Sheffield, so they evidently did not starve.

L. M. Rasmussen (ed.), *Human Guinea Pigs, by Kenneth Mellanby: A Reprint with Commentaries*, Philosophy and Medicine 134,
https://doi.org/10.1007/978-3-030-37697-0_10

At the outset we had taken over some allotments, mainly as a means of keeping everyone occupied in a healthy way. At one stage there was a possibility of developing this side of our activities. Among pacifists there are many who believe in "Community life" and communities have been started on the land with the idea of "returning to nature" and living on the produce of one's own labours. In wartime such ideas have had a special impetus, for some have felt that even in submitting to rationing schemes they are therefore supporting, or at least acquiescing in, the war economy of the State; others have felt that it is illogical to eat imported food, which can only be brought to Britain as a result of military or naval activity.

Some of our volunteers shared, to a minor extent, these aspirations for membership of such pacifist communities, and, although the experimental work would obviously have made such an ideal impossible in Sheffield, it was thought that it might be practicable to get some agricultural experience and, at any rate, contribute materially to the upkeep of the Institute. In addition to our allotments we therefore added a dozen hens and a number of goats to the establishment, and produced vegetables, eggs and milk for consumption in the house. On occasion, some of the volunteers did a really good job of work on the allotments, but on the whole it was apparent that it was just as well that we were not entirely dependent for food on the labours of the volunteers. The animals, somewhat surprisingly, survived, and the allotments produced some crops, but at no time was our agriculture a paying proposition. Our standard of work was presumably about the same as that of most pacifist communities—when the labour situation of the household was rationalised one man took over the responsibility of the hens, goats and a few rabbits and called this a full-time job, though it involved a good deal less effort than the outdoor tasks of any farmer's wife, who also has to look after a large farmhouse, a family of children and probably the feeding of the farm staff. But then, farmers' wives seldom have much time for spiritual meditation!

The second grant from the Ministry of Health expired in the spring of 1942. It had been a somewhat new departure for the Ministry itself to give a direct monetary grant in aid of an investigation of this kind. One purpose of the Medical Research Council is to support research in which another Government Department is interested, and on the recommendation of the Ministry the Medical Research Council now made us a grant to continue our work for a further period. During the time when we were closely associated with the Ministry we had always collaborated in a satisfactory manner. I had kept them informed of all developments and they had given me a completely free hand with the scientific work. This freedom of action was a very satisfactory feature; I now realise that I must owe a great deal to one or two medical and scientific people who would naturally be consulted by the Ministry and asked to advise on the conduct of the work. Fortunately, those concerned had themselves a good deal of research experience and knew that work like this must be run on the spot with a minimum of administrative control.

When we first planned the scabies work we told the volunteers that they would be infected in various ways (for we were concerned mainly with transmission), but it was understood that as a rule when infection had occurred they would almost immediately be cured. I now found that I wished to follow the course of infections which continued for much longer periods. When this was explained to the volunteers they all expressed complete willingness to submit to long periods of infection, although this went far beyond the original experiment. Some of them now suffered from scabies continually for as long as nine months at a time. They were regularly inspected during the whole of the period and an accurate estimate of the parasite population which caused the disease was made. This experiment was very unpleasant for the participants. They had, many of them, in the first experiments been infected for a few weeks only and they felt rather that the symptoms of "intolerable irritation" and other unpleasant experiences attributed in the literature to clinical scabies tended to be exaggerated. They soon changed their minds. After being infected for about a hundred days they mostly agreed that what they had previously experienced was negligible. Some kept rough brushes to rub over the skin to relieve the irritation. On cold nights some would rise from a sleepless bed and walk naked through the house, as when the skin was chilled the itching temporarily subsided and sometimes, if sufficiently tired, it was possible to fall asleep before the skin got warm and the irritation returned. Certain volunteers were reduced to sleeping naked as they scratched so vigorously in their sleep that their pyjamas were torn to shreds. Unfortunately, this constant scratching not infrequently damaged the skin and allowed the entry of sepsis-producing bacteria; these gave rise to unpleasant secondary infection, which could not be cleared up until the scabies was cured. I had sometimes to terminate infections as I was afraid that permanent damage might be done to the volunteers, though they themselves were quite willing to continue with the experiment.

Even now there are many people who think that scabies is a "dirt disease" and that reasonable personal cleanliness will confer an almost complete immunity. This view has been considerably shaken by the occurrence in recent years of the disease in individuals of irreproachable personal hygiene, but it still persists. We performed some experiments to test out this hypothesis. Volunteers subject to possible infection were divided into two groups, one group bathing daily and the other group being instructed not to bath at all and to wash as little as possible. The course of the disease was watched in both groups; towards the end of the second month examining the "unwashed" became a slightly unpleasant task! But the result of this experiment was quite clear cut, for both the clean and the dirty developed the disease equally rapidly and with precisely the same type of symptoms. The only difference was in very longstanding cases of the disease, where an individual with a dirty skin was more likely to develop secondary infection. The reason why the dirtier section of the community is more liable to scabies than the cleanly is presumably because they live under more crowded conditions, so that transmission is greatly facilitated.

In May 1942 the investigation attained the dignity of mention in the House of Commons. On the 14th of the month Mr. C. H. Wilson, the Labour Member for the Attercliffe Division of Sheffield, asked Mr. Ernest Brown, the Minister of Health, "whether he can give any information regarding the investigation of scabies by Dr. Kenneth Mellanby; how many persons were infected with the disease; how long the period of examination lasted; what results were obtained; how many of the persons infected were conscientious objectors; whether it was a condition of registration that they should undertake this experiment; and whether there were any temporary or permanent dangers attaching to it?" (This and the following quotations are reproduced from "Hansard".)

Mr. Ernest Brown then replied: "This investigation began at my instance, but is now being continued under the auspices of the Medical Research Council. Its object has been to obtain more knowledge of the parasite, the means by which scabies is spread, and the best methods of treatment. For this the services of human volunteers were essential. Forty-seven volunteers have been infected. They have remained under examination for varying periods up to eighteen months. The investigation is still in progress, but preliminary results are to be found in the White paper now available regarding scabies. Of the volunteers infected, the majority (and all who are infected for more than one week) are conscientious objectors; none of them has been directed to this duty by a Tribunal, though some who were already engaged in it when they went before a Tribunal were exempted from military service on condition that they continued in the work or took up first-aid duties. They have suffered discomfort rather than danger, but I am advised that they have not been free from risk of some prolonged disability."

Mr. Wilson then added: "Was there a medical examination before these people underwent this experiment, and has there been a medical examination since and with what result?"

To which further question the Minister replied, somewhat irrelevantly if otherwise gratifyingly: "The matter comes at the moment under the Medical Research Council, and the examination is still continuing. It is most valuable work, and I have no doubt it will have valuable results."

These questions and answers may require some elucidation. Mr. Wilson, who is one of the oldest and most respected Members of Parliament, has himself pacifist views, and has always been on the lookout for cases of the victimisation of conscientious objectors. When he first heard of our work he imagined, as did many others, that the volunteers had been directed to take part in the investigation and that some element of compulsion had been used to make them submit to unpleasant or dangerous experiences. A short time later we had the privilege of a visit from him at the Institute, and he was able to learn the exact position of the men from their own mouths. He was later able to help us in a number of ways.

Round about this date we also had to deal with the "celebrated case of Richard Henry Wodeman". This volunteer had been with us for some time, but his position was a little anomalous. I have something further to say about the registration of conscientious objectors in a later chapter, and will not refer to this matter in detail at this stage. Sufficient to relate that Dick, for reasons best known to himself, did not recognise the right of any Tribunal to judge his or anyone else's conscience, and he refused to appear before the body appointed to decide whether his objections to military service were truly conscientious ones. The Tribunal took the only step which it felt open, stating that in his absence it could not give him exemption from military service on conscientious grounds if these grounds were not specified, and his name was therefore put on the military register, i.e., he became liable for call-up for the Forces in the usual manner. Some time after this Dick was summoned for his army medical examination and he failed to attend. After a few more abortive invitations of this kind he was instructed to appear at the local magistrates' court, which ordered him to pay a fine of £5 and also submit to medical examination. As he still refused he was committed for trial for failing to carry out the orders of the court.

The next day Dick appeared before the magistrates once more. They felt themselves to be in a difficult position and, in the words of a headline in the "Sheffield Telegraph", "C.O. Praised by J.P. Who Jailed Him." The chairman stated that he was sure Dick was doing useful work and the prosecuting lawyer gave what was, in effect, an address for the defence. Now the law is such that a man who states that he is a conscientious objector, but who has been found by a Tribunal not to have sincere convictions to that effect (Dick was in this position due to his non-appearance at the Tribunal), may appeal for a further hearing if sentenced for a term of imprisonment of not less than three months if his offence is committed on allegedly conscientious grounds. A shorter sentence carries no right of appeal. It was clear that the magistrates did not wish to sentence Dick, but in order to allow him to appeal they gave him the minimum possible sentence of three months' imprisonment. The three other "conscientious" objectors who appeared on a similar charge at the same time were all sentenced to twelve months' imprisonment.

Now that Dick was in gaol there was a surprising outburst calling for his release. Many newspapers, including those unlikely to be suspected of tender feelings for "conchies", came out with very spirited protests, as, for instance, the following leader from "Reynolds":

> "CONSCIENCE. Richard Wodeman, for a soldier's pay, does the job of a human guinea pig. He has been inoculated with dangerous germs. He denies himself food and water as part of an experiment to aid shipwrecked sailors. He is a hero with a conscience which commands respect.
>
> "Unfortunately, the military authorities do not share the public's respect for conscience. Because Wodeman objects to military service he has been sentenced to three months' imprisonment in defiance of his record and of the Government's clear pledge that such men will not be the victims of stupid officialdom.
>
> "This sentence should be quashed."

Dick's case was also heard in Parliament, where Mr. Wilson asked the Minister of Labour "Why a man who was undergoing an experiment for alleviating sufferings was arrested during the course of the experiment and prosecuted; whether it is intended to prosecute others of the 47 men who have been undergoing the experiment; and whether he can make any statement about Richard Wodeman, who was sentenced by Sheffield Court to three months' imprisonment for refusing to undergo medical examination."

I must say I think the authorities were in a most awkward position and it is difficult to see how, with the best will in the world, and unless they were willing to accept the contention that no Tribunal had any right to judge anyone's conscience, they could have acted in any other way. The Home Secretary was appealed to by various individuals and bodies, which I imagine was somewhat embarrassing for a man who had himself at one time been identified with pacifist views. Dick went off to prison and was encouraged to appeal so that his objections to military service could be further examined, but he refused to take this step. After a few weeks of imprisonment the Home Secretary then intervened and he was released from gaol and returned to us in Sheffield.

The various questions and answers in Parliament mention 47 volunteers. This includes the men resident at the Institute, who were infected for such long periods. It also includes a number of other pacifist friends of the volunteers, who were mostly infected on restricted parts of the body only, although accidental escapes of mites caused a more general and lasting infection in a number of cases. Finally, some of the army medical officers who came for instruction submitted to infection for limited periods in order to become fully familiar with the parasite from personal knowledge!

In addition to being infected and working at the treatment centre, the volunteers did a lot of useful "field work" connected with the scabies investigation. We wished not only to be able to cure scabies when it was sent to a treatment centre for attention; we also desired to learn whether it was possible to do anything to prevent infection from ever taking place, or from "getting a hold" if the primary infection could not be prevented. To investigate a point of this nature it was necessary to find some more or less static section of the population where scabies existed and see whether there was anything to do to reduce the incidence of the disease.

For work of this kind the first essential is a thorough familiarity with scabies in all its forms; the volunteers certainly possessed this qualification from the most intimate experience! They could all diagnose scabies and check their diagnosis by finding and removing the causative parasite and demonstrating it under the microscope. Now we had already discovered that the majority of itch mites in any case of scabies were to be found in the skin of the hands and wrists, and a simple rubbing of these areas with benzyl benzoate emulsion (the most efficient substance for treating scabies) will cause their death. Treating of these parts of the body only is not recommended as a method for the cure of clinical scabies, but we thought it might serve to kill parasites which had recently attacked a new patient before the clinical disease had developed. We planned to determine the incidence of scabies among a considerable number of individuals, institute regular anointment of the hands and wrists and see the effect on the commonness of the disease.

It was impossible to carry out the experiment with any army unit, as military necessity causes so much movement of men that it seemed unlikely that the same individuals would be available at the end of the period of any experiment. We tried to work with the W.A.A.F. members of balloon barrage crews, but even they were so frequently transferred from one site to another that the results were inconclusive. In the end the investigation was carried out among school children. We obtained permission to visit schools under a number of different local authorities, and examine the children for scabies. Then the majority of the children had their hands and wrists treated with benzyl benzoate emulsion as already indicated and the examinations were continued. In all, well over a quarter of a million individual inspections were carried out, which is probably a record for this type of investigation.

It was a new experience for the volunteers to make these regular visits to schools and to examine the children. The investigation gave them a great deal of work, for the schools were sometimes at a distance and this necessitated very early rising and setting off long before the end of "black out" in the winter. The men soon developed a good professional manner, and carried out their work so well that they were generally welcomed by the staffs of the schools, although their visits caused a few minutes' interruption of the ordinary class work. We found the teaching staffs very co-operative, and, on the whole, the medical officers gave us a great deal of help, though occasionally, where our results indicated a possible break-down of their system, the implied criticism was somewhat resented.

I was very pleased to see how well the volunteers took to this work in schools. Although it meant early rising and a very strenuous time, they were very keen and never grumbled about being overworked. It indicated very clearly that they were willing to exert themselves very considerably at any job that they thought really worth while, even if they appeared lazy when put on to dull and apparently useless jobs about the house. These "prophylactic" experiments gave us a mass of very useful data, and we reduced the incidence of the disease very considerably. Our general conclusion was, however, that the home was the main source of infection among school children, and that no prophylactic scheme was likely to be entirely successful unless the whole family could be included.

With scabies so common throughout the country, many dermatologists and others were worried that the disease might never be controlled unless the medical and nursing professions could be brought properly up to date. Between about 1924 and 1935 scabies had been a comparatively rare disease, and many doctors and nurses trained during that period had little opportunity of becoming familiar with it. Dr. H. MacCormac urged that special arrangements should be made to give instruction on scabies, and, when it appeared impossible for this to be done on a sufficiently large scale by lectures and demonstrations, he suggested to the Ministry of Health and to the Army Medical Authorities that an instructional film for general practitioners, army medical officers and public health personnel might suit the purpose. Major Johnson and I had also considered this question independently, and had gone so far as to send a tentative script to the War Office.

The Public Relations Division of the Ministry of Health was very interested in the suggestion and asked the Ministry of Information to commission the film. A committee discussed the matter and suggestions for a script were put forward. I was summoned to a meeting and met Mr. Robin Carruthers of Spectator Films, who was to direct the film. He had a difficult task, for he had to fit in with the suggestions of a body of medical and scientific workers who by no means always agreed with one another; I think that the final result is greatly to his credit. I found that the committee was not at all sure just how much of the biology of the causative *Sarcoptes* it would be possible to present, and I did not know whether the mite would have a good "screen personality". Fortunately the micro-photography was to be entrusted to Mr. Frank Goodliffe of Science Films, who is probably the most experienced and successful worker in the country when it comes to subjects of this type. I had imagined that it would be possible to do the micro-photography only in Sheffield, where we had plenty of infected subjects and other biological material. Mr. Goodliffe and Mr. Carruthers came up to Sheffield to spy out the land, and I discovered that it was unlikely that they could, in war time, transport all the bulky and delicate apparatus required to Sheffield, and in the end I agreed to try to make the mites perform in Mr. Goodliffe's studio in Letchworth.

Two of the volunteers who at that time were infested with quite unusually large numbers of itch mites spent several days in the studio in Letchworth Garden City. I found it a most enjoyable business working with Mr. Goodliffe. It is always a pleasure to see someone really competent at work, and with comparatively little fuss, bother or waiting about (exceedingly little waiting by the usual standards of the cinema industry) he produced some first-class motion pictures of the various stages in the life history of the *Sarcoptes*. I found it refreshing to meet someone who knew just what he could and could not do—I would show him some feature regarding the mite and say what I thought should be made to appear, and at once he could say if it was possible, impossible or merely difficult—and if the last it was still carried out in a short time. The volunteers proved a never-ending source of biological material for the camera, and of entertainment to the staff of the studio, who had not previously dealt with "human guinea pigs". I was interested to note that the not inconsiderable expenses cheque paid to the volunteers for their services was equally distributed by them among the remainder of the men who had not been selected for this job. They felt that it was only chance that had made them so rich in parasites at the critical time, and so all should share equally as all had suffered to the same extent. I imagine that under similar circumstances I should have pocketed the money!

There was a good deal more to be photographed than the life history of *Sarcoptes*, and the rest of the work was done in various places in London—another studio, a hospital clinic, a disinfesting station and so forth. One volunteer served as a "model" for the clinical pictures after his infection had been accentuated with grease paint by Dr. MacCormac. I found it very interesting to learn just how a film was built up, though the short periods of hectic activity followed by apparently interminable periods of inaction were quite infuriating; this method of working seems inevitable in the cinema. All the time there was a good deal of gentle disagreement between the experts, which meant that the script was seldom the same for two days at a time, so that the unfortunate Mr. Carruthers must often have wished that we were all heavily infected!

Eventually all the pictures were taken, most of them had been scrapped, and the rest had been put together in a way which showed much credit to the discretion of the director and gave satisfaction—nearly—to most of the experts. I think we mostly felt that if we could start all over again we could make something which might almost justify the epithets bestowed by Hollywood on their more mediocre productions. So far the film was mainly silent, with a certain amount of synchronised sound for parts in which several of our more eminent medical authorities express themselves about scabies. The spoken commentary for the rest of the film was entrusted to Mr. Alvar Lidell. I must say I was impressed by the way he did the job. He knew nothing about scabies and the film was full of terms he had never heard before. I was present at the recording to ensure correct pronunciation of these technical words, and the correct emphasis at certain points. I was amazed at the speed with which Mr. Lidell seemed to grasp it all, and after one short rehearsal he recorded the whole thing without a fault. Shortly after this Mr. Lidell left the B.B.C. for active service, and his voice was not heard giving the News, but I heard his film commentary so many times that when he returned to the Radio once more I expected him at any moment to relapse into the familiar "There's something on the needle … it's a mite … etc."

The scabies film seems to have proved a considerable success. It is only intended for doctors and nurses and others who may have to find or treat scabies, but it has been widely shown to such audiences; in fact, I believe it has been more widely shown than any other medical instructional film in this country. Though we could now do it better I think it is quite a good film. The subject is one very suitable for treatment in this way, for you can show a lot of things, like the life history of the mite, in a most graphic way on the screen, though they are difficult or impossible to demonstrate in any other manner. The *Sarcoptes* makes an excellent "star".

Chapter 9
Some Results of the Scabies Investigation

When we started working on scabies, one of the first things that we wished to do was to make an accurate assessment of the problem which we were tackling. The information at our disposal showed that scabies in 1940 and 1941 was much commoner than a few years earlier, and this increase was usually attributed to "the war". It is true that Dr. A. M. H. Gray had published conclusive evidence that the disease, which was common in 1919, fell rapidly in incidence for several years, becoming surprisingly rare. After 1926 the incidence very gradually increased for some ten years, and then in the years immediately before this war the increase was accelerated, but even then it was nowhere near the 1919 level. Also, Dr. J. Alison Glover in the annual report of the School Medical Service, "The Health of the School Child", voiced his alarm at the increase in the cases of scabies among school children before the war. But notwithstanding these warnings the majority of people, including doctors, assumed that scabies was solely a war-time disease increased by war-time conditions and, as mentioned earlier, it was the dirty soldiers with their filthy blankets who were thought to be the main culprits. Blame was also given to air raids, causing "shelter life", which again allegedly spread the disease—and presumably caused this spread even in towns which were never bombed, or even for many months before the bombing began! These views even now have a surprisingly wide currency.

When working on the head louse I showed that the patients admitted into infectious disease hospitals formed a surprisingly accurate cross-section of the population, notwithstanding some weighting towards the poorer section of the population (and of course it is necessary to make adjustments due to the preponderance of children). If, say, 10%, of the boys of eight years old among the patients in a fever hospital are verminous, then the percentage among all boys of eight years in that

L. M. Rasmussen (ed.), *Human Guinea Pigs, by Kenneth Mellanby: A Reprint with Commentaries*, Philosophy and Medicine 134,
https://doi.org/10.1007/978-3-030-37697-0_11

city who are verminous will not be very far from ten. The accuracy of this method of estimating the incidence of lousiness may possibly be questioned when definite ratios like the above are given, but if lousiness were found to be twice as common in one age-group in 1939 as in 1938, then there is little doubt that in this section of the population vermin must be twice as common in the former as in the latter year.

The methods applicable to the estimation of headlice were also applicable to that of scabies. Previous experience had shown me which of a number of hospitals kept their records in an accurate and easily-abstracted form, and in which the staff were likely to co-operate in an investigation of this kind. Fortunately there were more such hospitals than we could possibly tackle at this stage, but I selected a number of suitable institutions. In these, a patient who was admitted for any infectious disease—scarlet fever, diphtheria, or whatever it might be—was also carefully examined during the period of treatment, and if scabies was also discovered it was treated and the fact entered on the case sheet. From the case sheets it was then possible to extract information regarding the proportion of patients with scabies in addition to the infectious disease which caused their entry into the hospital.

A number of us visited these hospitals and abstracted the necessary records; I also took the opportunity of checking the way in which diagnosis was made and recorded, to ensure that our data were accurate. The results were quite striking, for they showed that in all the cities concerned a steep rise in the incidence of scabies had started in 1936 and continued until 1941 (at which time this survey was made); in most cities it continued to rise after 1941. It was clear that the war-time increase was only the continuation of a trend which was apparent several years earlier.

So far no completely satisfactory explanation of this increase in the incidence of scabies has been forthcoming. It was clearly not due to anything in the nature of increased overcrowding or poverty, for the rise started when slum clearance was making rapid progress and unemployment was on the decrease. As scabies may be transmitted venereally it has been suggested, with little supporting evidence, that there was an increase in moral laxity in the middle thirties. I suggested that dancing and cinema-going might be partly responsible, and someone wrote to say that he thought that "the comparatively distant contacts in the cinema were unlikely to spread the disease". With this I fully agreed, but pointed out that I did not refer to such distant contacts, and suggested that he should take a quick look round any picture palace when the lights went up unexpectedly during the "big picture"!

Incidentally most attempts to explain the increase in scabies have stressed some possible change in the habits of the population rather than in their susceptibility to infection. As a rule, any disease which shews a marked periodicity does so because it confers some degree of immunity on those who have suffered from it and then recovered; a further epidemic is unlikely until a considerable proportion of non-immune individuals has grown up or entered the population. Now all records from both recent and mediaeval literature suggest that scabies has always been a disease which has come in waves, and therefore it might be expected from this information alone to give rise to further immunity in its sufferers. However, general experience is such that no such immunity has been suspected, for many cases of chronic scabies, continuing over years, have been reported.

Our work with volunteers perhaps throws some light on this subject. We found that no absolute immunity exists, but that nevertheless it is much easier to infect a person with scabies for the first time than to re-infect him again after his original infection has been cured. During the long incubation period which we found for these first infections, the parasite "gets a hold" without causing any symptoms and without therefore stimulating the host to any defence measures. When the infection progresses eventually, in perhaps two months, to the stage of typical clinical scabies, the host starts to fight against the infection. Although scratching damages the skin and causes secondary infection, it is a defence measure and the finger nails remove many of the invaders from their burrows. Also at this stage the skin sometimes alters its properties locally and produces conditions unfavourable for the itch mite. These reactions on the part of the host are not sufficient usually to throw off the infection completely, but they reduce the number of parasites substantially. Now the more mites there are in a patient's skin, the more likely it is that he will transfer one or more to someone else and so spread the disease. We found that in volunteers after a very long period of infection few mites could exist, though clinical symptoms were widespread and we had what was usually considered a "serious", and therefore most infectious, case of scabies, though in fact such a case is of comparatively little danger to the community. It is in the population infected for the first time that large mite populations are produced and "carriers" tend to occur. Where scabies has once been rife and is "got under" it may smoulder on, but it is unlikely to be epidemic until there is a substantial population which has never known infection. I believe that this information goes quite a long way to explain the periodic nature of scabies epidemics, though there is undoubtedly more knowledge required before we have a complete and satisfactory explanation.

Our work in the hospitals described above enabled us to state quite definitely that the disease had been increasing rapidly for several years, and that the increase was quite independent of the war. The idea that scabies is spread in air raid shelters does not stand scrutiny, for the increase which certainly continued during the war was as marked in those cities which had no population living in communal shelters and little air activity as in places which had large and almost permanent shelter populations. From my own observations I should think it would be very difficult to contract scabies in any of the larger and better controlled air raid shelters!

As well as determining this trend in the incidence of the disease, we were able to make an estimate of the actual incidence of scabies in a number of cities. Later it was possible to carry out actual examinations of nearly a third of one of these cities' 60,000 children, and we found that our original estimate based on the fever hospital findings was surprisingly accurate. The difference was only in the order of half per cent. This, incidentally, supports the view that the earlier louse data were pretty near the truth.

Taking the country as a whole, we decided that our information suggested that about the end of 1941 between one and two million persons must be infected with scabies throughout Britain. It was indeed a serious problem. When studying the extent of the infestation we paid a good deal of attention to the army, for scabies was reported to be rife there and soldiers were largely blamed for the rise in incidence, even after it had been demonstrated that the disease was already on the increase in peace time.

There was undoubtedly a great deal of scabies in the army, but it was soon apparent that there was, if anything, less than among the civil population. The explanation for the prevalent view was that the Forces was the only section of the population subject to frequent and regular medical inspections, and if ever a soldier, sailor or airman developed scabies it was pretty sure to be detected. Among civilians, however, the disease could flourish quite unrecognised. We found evidence to suggest that though soldiers certainly brought scabies into the army they seldom, if ever, passed it on to their comrades. It was exceptional for a soldier to contract the disease in the course of his normal military duties; infection took place from civilian sources, on leave or elsewhere.

The experiments on transmission established quite clearly that close personal contact was the commonest way of contracting scabies, for under these circumstances mites were most likely to be transferred from one person to another. Transmission through the intermediary of inanimate objects such as bedding was shown to be uncommon, even if it were occasionally possible. Where young adult males are concerned the most usual means of catching scabies is venereal, which in this case means sharing a bed with an infected woman, rather than the more casual and brief contacts which so often spread syphilis and gonorrhoea. At one stage of the investigations our findings were included in a play in verse about scabies and its social implications, produced at a Christmas party; the question of transmission is briefly dealt with thus:—

"Recondite research on *Sarcoptes*
Has revealed that infections begin
On leave with your wives or your children
Or when you are living in sin,
Except in the case of the clergy
Who accomplish remarkable feats
And catch scabies and crabs
From door handles and cabs
And from blankets and lavatory seats."

Being familiar with this rhyme a good deal of amusement was caused in the household when we read a paragraph in which a Medical Officer of Health in a Western county was stated to have informed some committee that the increase in scabies in that area was due to infection from "lavatories". The general concensus of opinion was that there must be a great number of parsons in the county concerned.

Once when lecturing to a military audience I said that scabies was usually contracted by "picking up a young adult female". When order had been restored I had to point out that I meant a young adult female *Sarcoptes*, but that the audience was probably right as well.

Now bedding could have been of importance in scabies infection in two different ways. First it could be responsible for infecting new cases, though it would not be likely to be of major importance in this way, for when a strange bed is occupied without clean bed-linen it is usual for there to be a strange human occupant present also—and this will certainly favour transmission. Of possibly greater importance is the reinfection of people after treatment—the theory is that they themselves are cured, but relapse due to itch mites left behind in their beds. This question of the importance of disinfestation of bedding and clothing has caused a great deal of trouble, and there are still some who attach great importance to it. The idea that you need not disinfest the bedding, etc., of cases of scabies is not a new one, for Hebra dealt thoroughly with this question a hundred years ago, and gave good evidence that it was unnecessary.

One heard more arguments about disinfestation. One Medical Officer would say that he got excellent results at his treatment centre, where he gave such and such a treatment, and in addition he disinfested all clothing and bedding, which was, he was sure, a necessary step. Then another would say that he got good results with exactly the same treatment and no disinfestation whatever. This at first sight may appear controversial, but should have been conclusive, for if anyone was getting uniformly good results without disinfestation then the process must surely have been irrelevant.

When we began to feel that all this "stoving" of bedding was largely a waste of time, we carried out some direct experiments. Several hundred patients were given the identical treatment for scabies (a single thorough painting with benzyl benzoate emulsion) and every alternate man had all his belongings disinfested while the remainder received no such treatment. The result was identical—under 1%. of cases in either group relapsed. It now seemed time to get some official pronouncement on the subject. A meeting of the Ministry of Health's advisory committee was summoned, and the evidence laid before it. Two Medical Officers of Health, Dr. J. L. Burn from Salford and Dr. J. R. Graham of Chesterfield Rural District, both described the satisfactory results which they had obtained after discontinuing disinfestation. And Lieutenant-Colonel M. H. Brown, Director of Hygiene of the Canadian Forces in Britain, described similar findings among his troops. As a result the Ministry issued a revised version of its Memorandum on Scabies, to this effect: "Methods of Spread and Prevention. … Recent investigations have shown that whilst articles of clothing and bedding may play some part in the spread of the disease the importance of these as a means of dissemination has been much overestimated: and that with the standard methods of treatment (benzyl benzoate emulsion or sulphur ointment) routine disinfestation of clothing is unnecessary, as the layer of medicament which covers the body after treatment is usually sufficient to kill all mites on the clothing. Moreover, in view of the small reinfestation or relapse rate which has occurred when disinfestation has been abandoned, it is simpler to repeat treatment in such cases rather than to expend the amount of manpower and material required for routine disinfestation in all cases. Emphasis should rather be laid on the follow-up and treatment of the family contacts and other close contacts of the original case, including, for instance, children of other families who are playmates with an infected child. It is not necessary, therefore, to insist on disinfestation as a routine procedure."

More recent experience has entirely justified this pronouncement. In every treatment centre where treatment is properly carried out, the abolition of disinfestation has had no ill effects on the relapse rate, rather the contrary, for the staff now has time to concentrate on its more essential tasks. It is impossible as a rule to evaluate the results of research in pounds, shillings and pence, but these experiments which were largely responsible for the abolition of "stoving" can easily be shown to have paid for themselves many times over. Dr. Graham has shown that in his area alone the ratepayers have been saved an annual sum of about £1000; this is in a population of some 100,000, so the potential saving in the whole of the country is in the region of half a million pounds annually. The cost of the research was about £5000 in cash, and a great deal of time and discomfort on the part of the volunteers. I hope their contribution will not be forgotten by the ratepayers![1]

Many have thought that scabies should be made a notifiable disease, that is, when a case occurs its reporting to the Public Health Authorities should be obligatory. One advantage of notification is that it may allow accurate statistics of the incidence of a disease to be compiled. I say "may" advisedly, because the statistics will only be accurate if the diagnosis is correct, and also if the data are dealt with honestly. I have unfortunately heard of Medical Officers of Health, no doubt with one eye on a lay Health Committee which expects everything to be better and better each year, and blames any deterioration on the M.O.H., recording deaths which should be attributed to, say, dysentery as "heart failure", a diagnosis which may possibly be accurate as to the final cause of death, but which is nevertheless most misleading. Incidentally, as a whole, one can say that the worse the statistics for any health problem in a city appears to be, the more efficient is the Medical Officer of Health, for it probably means that he is simply discovering disease that a less efficient man would leave uncovered. No wonder smooth-tongued and inefficient M.O's.H. are the most popular with local councils!

If a disease is notifiable, the Public Health Authorities must be informed of its existence, after which the treatment becomes their responsibility. It would therefore be impracticable to make scabies notifiable where no facilities existed for its proper treatment. Again, there are local authorities who have no one on their staff competent to diagnose scabies accurately, so they would find it difficult to control. When the question of notification was brought up in Parliament mainly as a result of agitation by those who really could deal with the disease in their areas, the Minister was at first most discouraging. Later policy changed and the "Scabies Order (1941)" was introduced; this comes under the Defence of the Realm Act, and gives Medical Officers of Health somewhat nebulous powers to inspect people suspected of scabies, and to compel their treatment if this is not carried out voluntarily. The Scabies Order is a half-way step to notification, and when it was introduced many felt it would be useless and that it would be impossible to secure a conviction of a recalcitrant patient. In fact the Order has been quite useful; there have been a number of

[1]The figure of £1000 for a centre and £500,000 for the country as a whole was based on incomplete costings made in 1943. Even at 1943 costs it should have been multiplied by four or five, to allow for the cost of premises, transport and administration. In 1973 the costs would obviously be very much greater.

successful prosecutions and the threat of possible prosecution has usually been enough to ensure treatment.

Even when the Scabies Order was introduced, the campaign for making scabies notifiable did not cease, and eventually a compromise solution was reached. In September 1942 the Minister of Health informed Parliament that "while it would not be appropriate to make scabies notifiable generally, he was prepared to consider applications from local authorities to have scabies made notifiable in their district, upon being satisfied that the Authority concerned had adequate facilities for diagnosis and treatment." Now the disease has been made notifiable in those cities where it is dealt with efficiently; the disadvantage of the system is that there is no way of compelling inefficient authorities to provide proper facilities, and there are still many areas without any treatment facilities whatever.

So far I have not said very much about the treatment and cure of scabies. It would be very difficult for anyone to develop an entirely new method of treatment, for almost every conceivable substance under the sun has been used. What we set out to do was to try the various remedies and see just how efficient they were, not only at relieving clinical symptoms, but also as substances which killed the actual parasitic *Sarcoptes.* I have already described the use of sulphur ointment, possibly the oldest remedy, and still one of proved efficiency. It was when this method was in use that the following verses, describing four common methods (the second and third have since been almost entirely abandoned) were composed:—

The treatment most in vogue of late
Concerns the benzyl benzoate,
Made up in water it's applied
All over every patient's hide.
It cures a very high proportion
When used with care and skill and caution.
Derris root yields rotenone
Which some say will cure alone
Without a bath or scrub or soap
(This surely is the soldier's hope)
This treatment seldom has a failure
But woe betide the genitalia.[2]
If hypo first is sprayed on thick
Then HCl may do the trick
Sulphur produced in nascent form
May kill *Sarcoptes* if they swarm.
Apply the liquids many a time,
And finish off with calamine.
But if you must be safe and sure,
Use the good old-fashioned cure.
Cover the victim with soft soap
Then in hot water let him soak.
A scrubbing next is his appointment
And finish off with sulphur ointment.

[2] Derris produces a severe scrotal dermatitis.

Actually, though there is nothing incorrect in these verses, our emphasis has rather changed and we prefer to use benzyl benzoate to sulphur ointment, especially where efficient and intelligent orderlies carry out the treatment. We have found that so long as the whole of the skin surface of the body, except for the head, is painted with a layer of benzyl benzoate emulsion in water, a complete cure will be obtained in practically every case. Instead of the somewhat heroic treatment described in earlier chapters this simple painting is a great improvement. Experiments have shown that although bathing does no harm it is not essential, and half an hour is ample for the complete treatment of any ordinary case. One thing which may be difficult to ensure is the painting of the whole of the body-surface—so many people will confine their attention to "the areas which itch", and then relapses become the rule. Finally, good results will only be obtained where whole families are treated at one and the same time, including those who do not appear to be themselves affected, for it is very likely that they are harbouring a few mites and will shortly begin to show clinical symptoms.

When I started working on scabies I was a little worried as to what might be the attitude of the dermatological specialists to my investigations. One is always told that medicine is such a closely guarded profession, and that its practitioners are always suspicious and even hostile to outsiders who make some attack on its problems. It is therefore satisfactory to record the liberal attitude of all the leading dermatologists in this country. They were quite rightly often critical of any finding which might not conform to the views they already held, but they were always willing to discuss problems, and, when convinced, to adopt a new viewpoint. The investigation owes a very great deal to many helpful discussions with dermatologists, for their long practical experience often suggested new and fruitful lines for laboratory investigations.

Chapter 10
Some Further Experiments

As has been stated in an earlier chapter, I have always felt that the opportunities we had of using human volunteers for experimental purposes must be exploited to the utmost. I have described our first dietetic experiments; when they were completed we looked around for some further suitable investigations to pursue. All manner of projects were considered and in this chapter a few which were actually tackled are described.

During the spring of 1942, at the time when the financial responsibility for the investigations was transferred from the Ministry of Health to the Medical Research Council, we were visited at Sheffield by quite a number of scientific workers who were interested in carrying out experiments with our volunteers. It was once more understood that the primary purpose of the Institute and its inmates was for the study of scabies, and that any other work must be carried out without upsetting the parasitological investigations. This cut out quite a number of possible experiments, but a good many were still practicable.

The Medical Research Council had a Committee which was concerned with medical problems which affect persons who are shipwrecked, and it was suggested that we should carry out some experiments into problems in which they were interested. The volunteers felt that this was essentially a humanitarian problem, and were keen and willing to co-operate. I was myself interested in questions of "water balance", for that was in essence what we were going to investigate. Though I had not studied this subject before in man, my work for some years had been concerned with not dissimilar problems as they concerned insects, and it was therefore very interesting to see how the same factors operated in these two completely different forms of life.

L. M. Rasmussen (ed.), *Human Guinea Pigs, by Kenneth Mellanby: A Reprint with Commentaries*, Philosophy and Medicine 134,
https://doi.org/10.1007/978-3-030-37697-0_12

The problem which we were to tackle was quite simple; it was to find the minimum amount of water which would keep a man in normal health and prevent him from becoming in any way desiccated. As the results were to be related to conditions applicable to shipwreck survivors, the solid part of the diet was to consist of the foods supplied to life-boats; these had an additional advantage in that they were virtually free from water themselves (most "dry" foods are in fact mostly water when it comes to analysis) so that we could easily determine the exact amount of the fluid intake.

Our first experiment necessitated going from tea-time on Tuesday until breakfast on Saturday without any liquid at all. This was done to determine certain necessary data which were required before the next experiments on a strictly limited fluid intake. The food (lifeboat supplies) had been planned to give the maximum amount of energy with the minimum bulk; in addition it was not perishable. There were four items of food—ship's biscuits, pemmican, malted milk tablets and plain chocolate. Under ordinary circumstances all these foods are excellent and can be eaten with enjoyment. The biscuits are not the traditional "hard tack" for they are reasonably thin and go excellently with cheese and butter, but without fluid they are difficult to chew or to swallow. Pemmican is made from dried and pulverised meat. It was originally prepared by the North American Indian as a type of meat which could easily be stored. He used buffalo meat, hung in the sun to dry and then chewed by the squaws to a paste, spat out and dried. Under the right circumstances Red Indian's pemmican probably made a tasty meal; the modern South American variety, which is no doubt produced from choice steers under hygienic conditions, is not so palatable when one's mouth is parched. The malted milk tablets were the least unpopular item of the diet. The chocolate was of excellent quality and was eaten with avidity on the first day of the experiment, but afterwards most volunteeers could not face any type of chocolate for a considerable time.

It appeared to be not too difficult to do without water for this first period of nearly four days. On the afternoon of the first day of deprivation most volunteers felt rather off colour, with headaches and other symptoms, but by next morning these had passed off: Curiously enough, thirst was not the most prominent sensation. For instance, it was never so intense as after hurrying across the moors in the blazing sun and arriving at the "pub" to find it is past closing time, although all the volunteers got into considerable water-debt. Towards the end of the experiment the volunteers began to look a little odd—eyes were sunken and tongues got furred. The most consistent desire on the part of the men was for some really succulent food, and some of them even dreamed of pre-war meals! During these experiments the men weighed themselves at frequent intervals, all urine was measured and analysed and other tests were carried out. The loss in body weight was in the region of 10 lbs. or a stone. On the Saturday morning just before breakfast time the experiment finished. The last sample of urine was passed, the last body weight recorded. By this time thirst was very noticeable, but this may have been partly because everyone knew that he would be free from the experiment's restraints. A warning had been given against filling up too quickly with fluid, and it seemed best to start with hot tea which could not be drunk too quickly. All the volunteers spent the morning eating and drinking, finishing with an enormous dinner consisting of all the rations that had been saved during the experiment.

The volunteers then submitted to a number of similar periods of deprivation, except that instead of being allowed no fluid they took half a cup, a whole cup or slightly more water each day. The experimental procedure—weighing, urine analysis, etc., was the same in all the experiments. Between each test a number of days on an unrestricted diet were needed to replace the reserves which were exhausted during the experiment. The main result of the experiment was to show that small amounts of water, in the region of a tea cup full, have very little physiological value, though they may have some considerable psychological importance to those who have been shipwrecked. Our experiments suggested that to maintain normal health, and to avoid any sort of water-debt, a very large amount of water, at least a quart, is needed every day. Those who say that they never drink as much should bear in mind that vegetables, bread and other foods all contain very substantial amounts of water, in fact most people take in more fluid in their "dry" food than actually as drink.

When we were investigating the incidence of head lice in children, our results indicated that the proportion infected was very much higher than that found by the School Medical Service. It was suggested that this discrepancy was due to school doctors and nurses not having sufficient leisure to make a really thorough examination of the children's heads; at times they had to examine more than 50 individual children in an hour. To get further information on this point, we deliberately infected the heads of several of our volunteers with head lice which we collected from verminous children. Our volunteers were all males and their hair was reasonably short, but, even so, it was surprising how difficult it was to recover all the lice liberated; we never ended up with more than we started! We found that if half a dozen insects were present it might take several minutes of careful combing to find the first parasite, and quite a long period to recover them all. It was thus obvious that in a girl with long hair an inspection taking perhaps less than a minute, and including the examination of several different features, could only be expected to detect head lice if these were present in very great numbers. It is, therefore, not at all surprising that our data reveal a much higher incidence than that found by the school medical authorities.

The volunteers have also served as subjects for experiments on surgical shock. To quote from "The Physiological Basis of Medical Practice" by C. H. Best and N. B. Taylor (p. 416), "the state of general collapse which follows two, three or four hours after a severe tissue injury is variously spoken of as a *wound*, *traumatic*, *surgical* or *secondary* shock. It is a condition quite distinct from the so-called *primary shock* which supervenes immediately upon the receipt of an injury. Primary shock has a nervous basis, pain and psychic factors, through their effect on the vascular system, playing a prominent part. Secondary or surgical shock is characterised by a profound fall in blood pressure; pallor and coldness of the skin; cyanosis of the finger tips and lobes of the ears; sweating; fall in body temperature and of the metabolic rate; rapid shallow breathing; small, rapid pulse; apathy; and other manifestations of collapse." There have been many theories as to the cause of surgical shock, some workers considering it is due to poisoning as a result of substances produced in the injured tissue (the "toxemic theory of shock"). Others have thought that nervous factors play a predominant part. There has been a great deal of research on surgical shock, and recently Professor H. N. Green, of the University of Sheffield, and his colleagues had published results showing that they have isolated a substance which will, when injected into animals, produce many of the symptoms of this condition. In order that information on this subject should be obtained for human beings the volunteers have allowed themselves to be injected with Professor Green's preparation and other substances, in all manner of ways, and have submitted to many somewhat alarming experiences. These experiments are not complete at the date of writing, and as I am not responsible for them I cannot pass any further comments except that the volunteers are rendering a very useful service and bearing up well, although "a bad time is had by all".

For many years Sheffield has been known as a centre for work on the fat-soluble vitamins ("A" and "D"), and when we began to discuss dietetic experiments in which the volunteeers could play a part it seemed appropriate that these substances should be further investigated. As I had been connected with some work on vitamin A deficiency in rats, I wondered if similar experiments more closely allied to human problems could be carried out on the volunteers. I put forward this suggestion in correspondence early in 1941, before the wheatmeal bread experiment had been suggested; this bread experiment occupied about a year, but at the end of that period the vitamin A deficiency was again discussed, this time with more fruitful results.

In the spring of 1942, about the time the "shipwreck" experiment was in progress, the M.C.R. Vitamin A Sub-Committee heard that I was prepared to allow them some facilities for carrying out an experiment on the volunteers and decided that they would try to make a considerable investigation. It was felt that the dozen volunteers at present available were hardly enough and I therefore recruited a number of "part-timers", that is, people who worked in the town at various jobs, but were willing to eat only a special diet prescribed by the Committee, and also were willing to submit to the various necessary scientific tests.

In the first instance the vitamin A experiment, like the other dietetic work, was considered as subsidiary to the scabies investigation, but it developed in a somewhat unexpected way and for a time became the main reason for the existence of the Institute. This experiment is also still in progress, and as the results are being considered by the Committee I cannot say more about them. It may be added, however, that volunteers have put up with a very dull and restricted diet, with no breaks or periods of alleviation, for a stretch of practically two years, which must quite easily be the world's record for such an experiment. Even I, with my experience of what the volunteers will, and what they won't, do, have been surprised at their powers of endurance. As a rule one can get many people to submit to a sharp, sudden and very dangerous experiment, but considerably more credit is due to one who will go on month after month with a dull and unheroic dietetic experiment, which to the lay mind appears to be getting nowhere, even if he is told that it is yielding invaluable scientific results.

At one stage it became apparent that we had to choose whether we could continue with the scabies experiments, or whether the vitamin A investigation was to proceed; the two became, for various scientific reasons, incompatible. I decided that we had already progressed sufficiently with the scabies work to make it difficult to justify interrupting the dietetic research, so except for field trials the parasitological studies came to an end. This meant that the volunteers passed to a considerable extent out of my control, and became largely the responsibility of the Vitamin A Committee.

Under these circumstances the experiments have proceeded, if not always quite smoothly, at least without interruption and without a great deal of friction. I think that this reflects considerable credit on the Committee, and more on the volunteers. It took the Committee a little time to learn to handle the volunteers tactfully, and delays and apparent inconsistencies, which are inseparable from any work managed, not by one person, but by a Committee without a definite leader, caused minor upsets, but did not in any way disturb the scientific value of the work or of its results. The success also owes much to Dr. Krebs, who had already learned, by trial and error, how to manage the household. But I think that the most important factor was the tradition that the volunteers had built up in the comparatively short time for which they had been serving for all these experiments—a tradition to continue for as long as possible with any experiment for which they had offered their services.

I have sometimes felt, however, that there was something lacking about this vitamin A experiment. There was no one scientific worker whose wholetime job it was to deal with the volunteers, the experiment, and the various problems arising from the Institute. I had been in this position in the scabies work, and this had really run surprisingly smoothly. The members of the Committee all had a whole lot of other interests and activities, so that the volunteers were not their primary concern; there was no real foundation for such a belief, but at times the opinion arose that the Committee was not really bothered with the volunteers and their problems. For future work of this kind it seems essential that a whole-time worker should be responsible for the experiment, and should devote all his time and energy to it and to the men.

During the latter part of 1942, when I thought the Vitamin A experiment was drawing to a close (it was in fact only beginning), I began to think of experiments on malaria, and some experiments were proposed for using the volunteers for the more exact evaluation of new drugs. This proposal came to nothing, but a year later I found myself in a position to carry out this type of experiment. Unfortunately for me the majority of the men were in the thick of the dietetic work and it seemed unwise to complicate this issue by introducing malarial infection, although it was later possible to transfer some of our volunteers from Sheffield to another place, where several suffered from infection with malaria. I was concerned with a whole series of experiments in which we were trying to improve the method of administration of a drug, so as to prevent people from becoming infected with malaria. The majority of our volunteers were either members of the Friends' Ambulance Units or were recruited through the Pacifist Service Bureau, and one and all they gave excellent service. A substantial proportion developed clinical malaria and were unpleasantly ill, but they put up with this in a most praiseworthy manner. I shall hope to describe the services of these volunteers in more detail at a later date.

From time to time our volunteers have given their services in other short-term experiments. They have tried out new soap which it is hoped may prevent scabies infection, to see whether the medicament in the soap did any damage to the skin (it did not, fortunately). They have taken large doses of anti-malarial drugs under conditions simulating the dry tropics, and have thereupon been violently sick. They have served as "bait" for mosquitoes, and have allowed themselves to be bitten by large numbers of these blood-sucking creatures. In fact there was some justice in the statement of a senior Australian officer who said "he would hate to get into my clutches"!

Chapter 11
The Problem of the Conscientious Objector

So far this book has dealt with conscientious objectors merely as subjects for scientific experiments, and any comments on them as individuals have been incidental. In this chapter I propose to trespass on to more dangerous ground, and make a few comments on the volunteers in particular, and on conscientious objectors in general, as persons. My only qualification for taking this step is that I have probably seen more of conscientious objectors during the war than has anyone else who is not himself one of their number, and I think that I can take an objective view of them as individuals, and of their beliefs. I expect that what I have to say will be criticised as being too sympathetic, and also as being too critical. I am afraid that the pacifists who look upon themselves as seers and prophets of a new age, but who do very little that is practical to bring in the conditions they talk so much (so very, very much) about, will say that I have only a very superficial grasp of their problems; I would say that I understand these particular problems of theirs only too well!

I have often wondered just what it is that makes a man decide that he is a conscientious objector. If one reads accounts published in the newspapers of 1914–18 one is given the impression that the usual cause is simply cowardice, together with the inability to stand up to dangers and discomforts. I think that no view could be further from the truth. From what I have seen myself I should say that conscientious objectors usually possess just those virtues which they are assumed to lack, and possess them to a quite unusual extent. That does not mean, however, that they lack other vices!

An experienced psychologist could no doubt produce an interesting and erudite thesis on the psychology of conscientious objection; I am unfortunately not qualified to do this, but the following comments may perhaps help some psychologist to make a proper diagnosis. Any "classification" of pacifists is bound to be an over-

L. M. Rasmussen (ed.), *Human Guinea Pigs, by Kenneth Mellanby: A Reprint with Commentaries*, Philosophy and Medicine 134,
https://doi.org/10.1007/978-3-030-37697-0_13

simplification, for they are on the whole individualists, and it seems unusual to find two who do not differ considerably in their opinion on some apparently fundamental aspect of their beliefs. But I think that they can be divided into two main and broad categories.

The first main group of pacifists includes those who base their objections to war on religious grounds. I realise that to the psychologist the bare statement that a man is "religious" means that we have not got much further; he will want to know *why* a man is religious, and whether it is because he feels improperly adjusted to this world and, therefore, is attempting to live in a world of fantasy. I do not feel prepared to tackle this question; it seems to me that a man can be logically religious or irreligious, and I cannot take the matter any further into the realms of theology or metaphysics. Many of the religious objectors are subscribing members of orthodox and established religious communities, while others, though in sympathy with Christian beliefs and ethics, do not belong to any particular church. Some of these latter were for many years subscribers to some church, but have now given up their membership because the leaders of their church are supporting the war. I personally think that the "religious" objector is taking a perfectly logical attitude, in that he accepts certain apparently fundamental Christian teachings on the subject of violence, killing and loving one's enemies. To me it seems much more consistent to be a pacifist if you are a Christian than for a Christian to be willing to undertake combatant service, though the majority of sincere and practising Christians seem to be able to overcome this dilemma.

The second main group of pacifists includes a most heterogeneous collection of individuals. On the whole they are "aggressive" types, whose real objection is not so much to violence as to being in a position where at any moment they may be ordered about by someone for whom they have no real respect and be ordered to carry out some duty which they think futile or unnecessary. I think that this general "bloody-mindedness" is the fundamental characteristic of this group and that their pacifism is often an unconscious rationalisation. By this I do not wish to imply that they are not sincere in their views, for they are mostly quite unaware of the details of the mental processes by which they have reached their "pacifist" conclusions.

Many of my acquaintances who are not themselves pacifists, or sympathetic to pacifists, have agreed that my volunteers have rendered a splendid service, but they have insisted that I only get the "genuine" pacifist and that the majority of C.O.s are really leadswingers, unwilling to undertake military service for quite unworthy reasons. They suggest that most people who say they are "pacifists" refuse to join the army because of fear, the dislike of hardships and discomfort, and particularly because they wish to remain in a more lucrative civilian profession. All this has, in fact, a little truth in it. No doubt, when they registered as conscientious objectors, these people whom I have employed as volunteers did hold their views with sincerity. And there are some people who have registered as C.O.s for base reasons, but I think these are very rare. If a man really wants to keep out of the army he can do so (or, at any rate, in the earlier years of the war he could) without incurring the stigma of being called a "conchy". He could, for instance, obtain employment at a high wage in an armament factory, where he would constantly be lauded to the skies by our politicians for his patriotism, while he slacked at his bench. (I do not wish to imply that more than a small fraction of "war-workers" are slackers of this type; I know many would join the combatant forces if they were allowed, and that the majority do a great deal of hard work, but there is a percentage of slackers who use their "reserved occupation" as a cloak for all the vices usually attributed to the unfortunate "conchy".)

As most people know little about the way in which a conscientious objector becomes recognised by the State as belonging to that category, it may be of interest to describe the procedure. From time to time the details have been adjusted, but the broad outline has remained the same.

When conscription was first introduced before this war, a pledge was given in Parliament that allowance would be made for those whose consciences did not permit them to engage in military duties, and that no one would be prosecuted or imprisoned for his conscience's sake. It was felt that it was not practicable simply to rely on the unsupported statement of each individual (and I think that the majority of pacifists will agree with this view), so special Tribunals were set up to investigate the reasons given by every person who called himself a conscientious objector.

These Tribunals have met in most large centres of population. They are presided over by a County Court judge and include usually a trades union representative and, perhaps, a lawyer or a clergyman. The members of the Tribunal are supposed to take a fair and unbiased view of the men who appear before them and of the reasons for their action. As a whole, those who have appeared before Tribunals say that the Chairman (the County Court judge) runs the show and the other members tend to be "yes-men", but this is perhaps a biased view. Human nature being what it is, the Tribunals are by no means uniform in their attitude to those who appear before them and in the pacifist community some Tribunals have a "bad" and some a "good" reputation. Having heard the nonsense that many otherwise sensible pacifists put out to a Tribunal, I think that as a whole these bodies have acted surprisingly well and fairly. I can conceive of no more appalling task than sitting as a member of a Tribunal day after day!

Nevertheless, although Tribunals have, on the whole, done their best, and have usually made quite a success of a difficult task, I think they have, at times, descended from their usual level. Sometimes it has seemed to me that one particular Judge has been most unsympathetic to those appearing before him, and others have not been able to refrain from cheap humour hardly in keeping with the dignity of the Bench. Thus, when one of our former volunteers appeared before a certain Tribunal for revision of his conditions of registration he described how he had taken part in our experiments and had been paid 16/- a week. The chairman then said: "That was cheaper than a guinea-pig," and later laughingly remarked: "You were doing a poor guinea-pig out of a job." However, as the man won his appeal, no doubt the opportunity of making these wise cracks put the chairman into a good temper!

When a man appears before the Tribunal he has to prove to them that his objections to military service are indeed conscientious, and the Tribunal can then direct him to work which will not offend his scruples. A man who is a subscribing member of a religious denomination, and who can prove that he has been a regular attender at church services and so forth, and, best of all, who can get a parson to prove his statement and confirm that he has been a pacifist for many years, has little difficulty in proving his case. A birthright Quaker will almost inevitably obtain exemption. Others who say their objections are "religious", but have no objective evidence to prove their case, have a more difficult time, though if they can state their case clearly and show a knowledge of religious literature they may have no trouble. Unfortunately, there have been a number of cases of "religious" C.O.s who say their beliefs depend on the teaching of the Bible but who cannot give any evidence that they have even read a verse of that book! Although a trained lawyer could always tie up a person who appears before him at a Tribunal and make him give inconsistent answers, the majority of the Chairmen manage to avoid this temptation, and they treat with consideration the rather stupid and non-eloquent, but probably very sincere, type of man who needs help in stating his case. It is only the "Smart Alec" type who gets a thoroughly bad time; any attempt to "show off" before the Tribunal is fatal, though at times the Tribunal will decide that a man whom it obviously detests is, in fact, sincere and worthy of exemption.

The non-religious pacifist has a much more difficult task and is apt to be told that his objections are "rational" and not "conscientious". On the whole, the Americans manage this in the simplest way, for they only recognise religious grounds for pacifism, but we in this country have tried to make the whole thing as fair as possible. In the majority of cases we have succeeded.

If the Tribunal decides that a man is a genuine conscientious objector it must then fix the type of work he is to do. Some are given "unconditional exemption", which means that the Tribunal is so convinced of their sincerity that it can trust them to find the most useful employment. Others are directed to the Non-Combatant Corps of the army, to the R.A.M.C. (if it will have them), to Civil Defence, the land and so forth. The vast majority of those who appear before a Tribunal accept the conditions either willingly, or, at any rate, with resignation, and they do not appear before the public eye again.

But when a man is not satisfied with the decision of the Tribunal, he has the right to appeal, and he then states his case once more before an Appellate Tribunal. These bodies seem to be more carefully chosen than the local Tribunals and, as a whole, even those who have their appeals dismissed come away with the feeling that they have had a fair hearing. Also the experience at the local Tribunal may have taught them to put their case in a more acceptable manner!

The authorities gave a pledge that no one would be prosecuted for the sake of his conscientious beliefs, and they have, I think, done their very best to keep this pledge. It is only after a man has been declared a "fraud" by both these Tribunals that he is liable to be called up for the army. But a certain number of people who are declared to hold no true conscientious objections to military service still object to being called up, and a proportion of these are eventually found to have a conscience after all.

What happens is that the "ex-C.O." is summoned for his medical examination for the Forces. As a rule he attends, joins the army and becomes a normal soldier. But he may still object, whereupon, after several such invitations, he has to appear before the magistrates for failing to obey. The court then imposes a fine, usually of £5, and makes an order that he shall submit to medical examinations. Some carry out the ruling of the court; those who refuse appear once more on the charge of refusing to comply with the orders of the court. They are sentenced to prison for a period of up to a year. If the sentence is under three months the man has to go through with it. If over three months he has another chance of appealing and appears once more before a tribunal, perhaps this time to have his objections allowed to be truly "conscientious"; but before this he will have experienced prison conditions, maybe for several months.

Some men, after being struck off the register of conscientious objectors, decide to join the army and then once more find that they cannot, or will not, comply with military discipline. For this they are court-martialled. If a sentence of over three months is imposed the man may once more appeal and have another appearance before a Tribunal. There was much complaint because some military courts made a practice of imposing a sentence of just under three months, and some men who later were found to have genuine conscientious objections spent years in the army going in and out of detention barracks the whole time.

Because some men who call themselves pacifists have been sent to gaol repeatedly it is sometimes said that the authorities have gone back on their pledge and are also indulging in the old "cat and mouse" tactics. This charge can be refuted, because the only people to suffer this treatment are those that the Tribunal has decided are devoid of true scruples of conscience—but as some men have appeared before as many as five Tribunals without any success and have then received exemption from the sixth, the question is a very difficult one.

I have said a good deal about the difficulties and mistakes of the Tribunals, but it must be realised that in the vast majority of cases they do their job very well. I have several times been surprised by the wisdom of their decision—they have said that a man is deluding himself and has no real conscientious objection to military service. The man thinks it over and finds he agrees and he becomes a willing conscript. And as well as denying a conscientious objection to those found later to possess one, Tribunals also exempt men who should not receive such treatment. I know of many instances of those who were given exemption and who later voluntarily joined the services, most often going into the R.A.F. as "air-crew".

A mental specialist has told me of some cases which are of interest in connection with the problem of pacifism. He has met a number of soldiers who have broken down mentally and some of these have done so because they are, in fact, pacifists. They have sincerely held pacifist views, but they were afraid to stand up for their convictions and so allowed themselves to be conscripted. Inside the army their trouble has not been fear, but a feeling of guilt that they have betrayed a cause in which they believed, and this has in time caused a serious breakdown.

When talking about our co-operation with the army, I have stated that we always found soldiers and pacifists got on quite amicably together. It is rare to find anyone who has experience of active service who is violently antagonistic to the pacifist. Those who want to "shoot the lot" are usually elderly women or men who were too old to fight in the war of 1914–18. As a whole, the country is much more tolerant to the C.O. in this war than in the last. One hardly ever sees the intolerant cartoons that appeared then, and this, again, I think is because most of the artists who are working to-day were soldiers before. It is interesting to note that many of the older leaders of present-day pacifism were soldiers 30 years ago, and they had excellent military records, too. Then, again, many of the pacifists of 1914–18 are among the foremost supporters of the war on this occasion. It is dangerous to be dogmatic about one's views and what they may be a few years hence. Most pacifists would agree with this; only the very young ones think they have made their irrevocable decision for all time.

Politically, the majority of the C.O.s belong to the "left wing", though they are so "minority minded" that they seem unable to agree with the policy of any of the major parties. They have rather nebulous ideas about establishing a world on more or less Christian and Communist ideals, and they hold innumerable little meetings to save the world on these lines. I find that as a whole even my volunteers are very immature on these subjects and inclined to go on talking for years in the way very young undergraduates discuss the salvation of civilisation. It all seems so ineffectual; but perhaps I am just cynical and intolerant. But although I avoid discussions of these topics, I cannot help hearing a lot of what seems nonsense coming from people who are at the same time so sensible about their experimental work. Also, as a whole, the man with the most ambitious ideas about world-salvation is generally the least likely to clean the doorstep when it is dirty!

Chapter 12
Science and the Press

Not long after we started the original investigation certain papers decided that the volunteers were "news". I had been a little afraid of this possibility and hoped to keep our investigations out of the public eye; but this was not to be. In May 1941 one of the men had to appear before the Tribunal at Leeds and there were newspaper reporters present at the hearing. Every now and then there are tit-bits of news to be picked up at these Tribunals, and the comments in the papers are usually to the disadvantage of the pacifists, for publicity is generally given to the more scathing remarks of the Chairman when he appears to have found a "lead-swinging" individual who claims exemption on inadequate grounds. The attitude to this volunteer was, however, quite different. The papers attempted to magnify the horrors of his sufferings and the value of his services. It so happened that at the time he had not yet developed a positive infection, so that all his suffering had consisted of contact with none too cleanly clothing and bedding, but some accounts were printed showing him to be a veritable martyr to science.

The interest of the press on this occasion was fairly short-lived and we were allowed to sink back into the oblivion we deserved and desired. But a year or so later information about the "shipwreck" experiment somehow leaked out and we were practically besieged by reporters. All day long there came trunk calls from London and other centres and personal visits from news hounds of all types, including "glamour girls" who tried to extract information by the exercise of their charms. Photographers wished to obtain the most impossible pictures ("one of the volunteers taking his first drink for four days") and one camera man pursued me into the country and suddenly appeared over a wall with his infernal apparatus, only to escape by telling a sad story about what would happen if he returned empty-handed.

L. M. Rasmussen (ed.), *Human Guinea Pigs, by Kenneth Mellanby: A Reprint with Commentaries*, Philosophy and Medicine 134,
https://doi.org/10.1007/978-3-030-37697-0_14

Probably over the whole experiment we have not averaged less than two enquiries per week from some newspaper or another.

As a result, these experiments have received much more newspaper publicity than they deserve and much more than many more valuable and erudite researches have received. I have been criticised a good deal by my scientific colleagues on account of this newspaper publicity. The criticism has been of two sorts. Those of my colleagues who have never done anything which could possibly be of interest to any paper have accused me of deliberately seeking publicity and of courting the interest of the press. Others who have accomplished a good deal themselves and have, therefore, had experience of the attention of journalists, while condoling with me to some extent, have suggested that I have dealt unwisely with the situation and should have taken more steps to keep the investigations out of the newspapers.

There may be some substance in the second of these criticisms, but the situation has not developed quite in the way my colleagues have suggested. When one is working, not with dumb animals, but with human beings as one's experimental animals, it can be very difficult to keep the matter entirely in the dark. Although the volunteers themselves may "keep mum", their friends and relations are apt to indulge in "careless talk", and so the papers get wind of the experiments. Now once a reporter has some information like this he will probably write a story anyhow, and the best plan is to give him some accurate information and try to ensure that he puts out a reasonable statement.

I have found, contrary to many people's opinion, that the best way to avoid unpleasant publicity in the newspaper is not to hide away from the reporters, but to give them an interview and tell them the whole story with the fewest possible reservations. Then he can be told just which of the things he has been told may and may not be printed; I have found that if one takes a reporter into one's confidence he will not abuse it. The most objectionable paragraphs have always appeared as a result of trying to avoid the gentlemen of the press.

Now while it is always possible to get a square deal from a reporter who is given an interview, the one thing to avoid is making any statement or answering any question over the telephone. Unless one is very experienced one finds that one is saying all the wrong things, and then they appear in print in an even more garbled form. And personally I find there is a grave temptation to hoax a journalist and tell him some completely fictitious story for publication in his paper. This temptation must be resisted, even if the truth is so seldom stranger than fiction.

Nevertheless, with an effort, it might have been possible to keep these experiments completely out of the papers, but I am not sure that that would have been entirely desirable. Had it just been that they wanted to print an account of my work it would have been different, but what they wanted to do was to give some account of the services of the volunteers. Now nearly always when a paper gives information about "conchies" it is saying something derogatory, so it seemed only fair that when they wanted to give a certain amount of credit they should be allowed to do so, particularly when the credit was entirely deserved. And these newspaper stories did the men some amount of good, even reconciling one man to his future mother-in-law, who decided, after reading about what he had gone through, that a pacifist son-in-law was not quite such a bad thing after all.

The appearance of accounts of our experiments in the daily papers had one unexpected effect, and that was that it started off our "fan mail". Some people wrote simply in appreciation of the work of the volunteers. Others wished to be helpful and sent us old family recipes for the cure of "the itch"; some of these were domestic methods of producing sulphur ointment from lard and flowers of sulphur, and would, no doubt, be most efficacious. Other remedies were more bizarre (though not unlike some that one reads about in the medical journals), the prize going to what was said to be an old gipsy treatment, where you made a body-belt of flannel containing a mixture of honey and various chemicals and wore this all the time until "it fell off". We were not sure whether this treatment cured the disease, alleviated its symptoms or merely made you so unpopular that you never had any chance of contracting infection. We received letters and booklets on other subjects from men with the most heterodox views. One attributed his health to the fact that he always drank his own urine and another rubbed this liquid into his skin daily. One man attributed most ills to over-indulgence in sexual intercourse, and stated that it was dangerous to health to indulge on more than a single occasion in every TWO YEARS.

The whole question of the attitude of the press and of the public to scientific work requires some thought and some action. Scientific workers as a whole deplore the lack of knowledge and interest on the part of the public when it comes to matters of scientific progress. On the other hand, they do very little to enlighten the public about their work. It is true that the popular press has a tendency to stress the least important aspects of research. … I soon found that if a human guinea pig was a "gurl" she had twice the news-value of a man … but this is partly because so little proper information is available to the journalist. We do require some proper scientific "public relations" organisation which will discuss the publication of objectionable articles on scientific work by issuing information of a proper nature. I am sure that the majority of newspapers would prefer to print accurate information from a source of this kind instead of sending their reporters nosing around among the seamier side of scientific investigation.

Scientists as a whole do little to enlighten the world about their subject. There are now a few first-class workers who have produced excellent "popular" books, but although this may have swelled their bank balances it has not increased their scientific reputation, for they are often considered to be guilty of prostitution of their subject, instead of earning the gratitude of their colleagues for making a wider public sympathetic to scientific investigation. In the long run, it is the public which pays for scientific work to be done; surely, they then deserve to receive some more information about it.

Chapter 13
Human Guinea Pigs, Past and Future

At some stage or other almost every advance in medical science has depended on the use of a "human guinea pig". Very often, as in the early work on malaria and yellow fever, the scientific workers have been their own victims. At other times new remedies have been tried out in desperation on patients who could not possibly recover if given the current and accepted treatment. Both these methods have grave disadvantages. The man in charge of an experiment should be making objective observations, which he cannot do if he is also the subject of the experiment. Furthermore, he is probably too valuable to waste as a mere guinea pig. On the other hand, there are sometimes advantages in first-hand information regarding what an experiment actually involves to the participants; I myself have usually gone through with any experiment once before asking a volunteer to follow suit, but after that I have been the observer and not the guinea pig. The other method of working with desperately ill patients is not always satisfactory, for it prevents the remedies from being tried out early enough, that is at the stage when they might be most likely to prove their worth.

Of course, human beings have often before served voluntarily as subjects for medical experiments, but, on the whole, these tests have been no more unpleasant or dangerous than are the majority of animal experiments which cause so much heart-burning to our anti-vivisectionists. (It is not generally known that if you keep two animals and feed them on different but otherwise normal diets, and make observations on their growth, that counts as "vivisection" and goes to swell the figure of the "inhuman tortures" inflicted without anaesthetic; should we chloroform our dogs before giving them their dinner?) Medical students, particularly in America, have given very useful service in this way, and recently similar experiments have been carried out in universities and factories in this country. But the extent to which such volunteers can be used is limited, for, as a rule, they must be kept in normal health for the majority of the time.

L. M. Rasmussen (ed.), *Human Guinea Pigs, by Kenneth Mellanby: A Reprint with Commentaries*, Philosophy and Medicine 134,
https://doi.org/10.1007/978-3-030-37697-0_15

The largest and most successful series of human experiments that I know of was made by Dr. H. Lyndhurst Duke during the early thirties in Uganda. Dr. Duke was working on sleeping sickness (a disease only found in tropical Africa, caused by a microscopic parasite called the trypanosome in the blood and transmitted by the bite of the tsetse fly). He obtained a great deal of valuable information about the cause, cure, and particularly the prevention of this disease through the co-operation of African native volunteers. Dr. Duke had one great advantage over the majority of research workers who have studied tropical diseases in Africa in that he was probably the best linguist in the Uganda Protectorate, which meant that it was possible for him to obtain the full confidence of his volunteers by discussing the work and its problems with the men. If, like most scientists, he had to rely on "kitchen Swahili" as a pseudo-lingua-franca for conversation with Africans to whom Swahili was also a foreign language, his experiments would have been impossible. Seeing this work at first hand gave me many useful tips on the management of human experiments.

I have mentioned that it is difficult for scientists to take a properly objective view of experiments in which they themselves are participants. This is true when the experiments deal with questions of disease, but it may be even more important when working on factors in diet and nutrition. I know of some work in which a number of scientists were testing out a particular substance which was of the greatest nutritional value and which might have been used to improve the standard of nutrition in many parts of the world. The scientists were so impressed with the excellence of their substance that they found it actually improved the flavour of ordinary foods; a test on lay volunteers revealed that it could be nauseating rather than delicious when eaten without the sauce of faith!

The most original feature about the investigation described in this book is that for the first time conscientious objectors were selected as volunteers and that a special establishment was set up to carry out the experiments. When it had been established that this system would, in fact, work and added facilities to research which could not be obtained in any other way, a good many other scientific workers explored the possibility of setting up similar machinery. I was approached by quite a number of investigators. Some wished to carry out what was virtually straightforward war work, which I felt was not at all suited for obtaining the co-operation and help of pacifists. Other research workers, though attacking a more suitable sort of problem, were temperamentally unsuited to manage volunteers possessing heterodox views on the war (members of the "shoot the lot" school), and I felt they needed discouraging gently but firmly from proceeding with their suggestions. But on several occasions I have put other workers in touch with pacifist organisations and satisfactory co-operation has ensued.

In the United States a good many pacifists have served as subjects in medical experiments, particularly in work concerning dietetic deficiency. These experiments started some two years after our Institute had been established, but, as in this country, the Americans have learned that though a man may be unwilling to undertake military service he may be able and willing to give valuable help in other ways.

One disadvantage about pacifist volunteers, when they are used in experiments on disease, is that they are much too "tough". Whereas, when "browned off" by training and so forth, the ordinary soldier tends to report sick, with a description of the most alarming symptoms, when there is the least little thing wrong with him, pacifist volunteers, unless carefully watched, will carry on even when seriously ill. I hasten to add that this comment on the soldier does not refer to his behaviour when action is imminent; then he will carry on notwithstanding appalling conditions and all manner of upsets to his health. But I find that care must be taken to avoid overlooking early symptoms due to the pacifists' dislike of being thought "soft"; the difficulty is to impress the importance of noticing symptoms without making a volunteer into a temporary hypochondriac.

The war has made a great deal of research difficult or even impossible, but, as I hope I have shown, it has also given us opportunities for carrying out investigations which would have been impossible under other circumstances. I have often wondered whether it will be possible to continue to carry out this sort of work in peace-time. The whole problem hinges on one question: Will volunteers be available?

I do not think that there is any doubt that it would be possible to carry out many human experiments in peace-time, and, in fact, it might be easier in some ways to secure volunteers then than in war-time. I originally selected conscientious objectors for this work because they were the only section of the population not likely to be called up or compelled to leave the experiment due to military or industrial necessity; they also included sufficient people with a humanitarian outlook which made them prove such admirable subjects for our experiments. But though these pacifists have, in fact, served us so well, not even the most strenuous opponent of militarism would pretend that others who during war-time were "otherwise engaged" would not have served equally well.

But if this type of work is to be carried out in peace-time it will have to be put on rather a different basis from the present one. We could not expect young men to devote several years of their lives to being guinea pigs in return for what is, in effect, a mere pittance if they could be making progress in their ordinary careers. In war-time most men are wasting so many years in the army or in a job unconnected with their usual profession that this feature is not an added hardship for the pacifist volunteer.

There is one circumstance which might make pacifists the most likely and most easily available volunteers in peace-time, however, and that is the continuation of conscription. It is dangerous to prophesy, but I think that it is probable that there will be far more people with a conscientious objection to military service in a peace-time army than have ever been found when the war is being fought (if this is so it certainly gives the lie to the idea that the pacifist is, therefore, a coward). One piece of evidence in favour of this is that a far higher proportion of men registered as C.O.s during the period of "phoney war" before the fall of France in 1940 than during the time when hard fighting seemed more likely to be the lot of everyone in the Forces. If conscription continues I think there may be many who will prefer to act as subjects for experiments for the usual period of military training, and who would not expect to be economically better off than they would have been in the Forces.

Assuming that there will be men and women who may be willing to offer their services as volunteers for scientific experiments in the post-war world, some special organisation will be needed to make the best use of their services. The ideal plan would be to have a special Institute for the study of "Human Biology". This could be situated in one of the large country mansions which will no doubt be empty at the end of the war, and I can see no reason why this work should not be carried out in one of the pleasanter parts of the country, as, for instance, the Cotswolds; this would be reasonably accessible to Oxford, Birmingham, London and several other university towns. Such an Institute would require a small but carefully chosen permanent staff of medical and scientific workers and a permanent nucleus of volunteers. There will always be a small number of men who, with reasonable prospects for the future, will be willing to make a career of this type of work and who will be invaluable in teaching new subjects the necessary principles of human experiments. As a whole, the volunteers should be recruited on a short-term basis, for one particular experiment which may last a month or up to six months. I should think that during the summer there would be the largest number of possible volunteers, which would include undergraduates during the long vacations and others, who would find a period in the country, even as a guinea pig, a not unpleasant change from work in the city. It would be much easier to accommodate large numbers in the summer; they would only need temporary buildings of a kind that might be difficult to heat in cold weather.

Assuming that such an Institute came into being, some scientific work would be in progress all the time under the direction of the permanent staff. But it would also serve as a centre where scientists from all over the country, or the world, could come for a period to work on their own problems. In the same way in which marine botanists or zoologists who are marooned at inland universities go to the laboratories at Plymouth or Millport during vacations to work on their own problems, so human physiologists and medical workers could come to our Institute. We would have a more difficult task than the marine laboratories, for they only need to provide bench space and access to the life in the sea and we would need to provide human material. It would be necessary to get long notice from all workers, and somebody would have the difficult task of working out the various "priorities".

I am sure that the project as sketched above would be feasible, provided that the funds were available. The two questions which would need to be decided relate to finance and to control. An Institute for the study of human biology would not be an inexpensive proposition, though the money involved may seem very little when compared with the cost of the war. The question of control is more difficult; there seems no doubt that it would have to be in the hands of a committee of scientists, but their choice would present many problems, for they would not only require a considerable knowledge of the scientific problems involved, but also the ability to understand and manage the volunteers. In the long run, however, I think the success or otherwise of the venture would depend on the choice of the small nucleus of permanent volunteers, whose work it would be to see that the correct sort of tradition was maintained, so that newcomers felt they were entering an organisation where they could really be of use.

Chapter 14
The End of the Sorby Research Institute

The following pages of this book were written 30 years after Chaps. 1, 2, 3, 4, 5, 6, 7, 8, 9, 10, 11, 12, and 13. I have explained how the scabies work was brought to an end, and the experiment on vitamin A was started. I showed how I handed over responsibility to the Vitamin A Committee of the Medical Research Council (p. 131) and I said something about the routine of this and other dietetic experiments. The purpose of the new experiment was to discover the minimum amount of vitamin A required by man to maintain normal health, so that, in times of shortage, supplies could most efficiently be distributed. It was generally believed that vitamin shortages were likely to occur and might have serious harmful effects in war-time Britain. Carotene, the chemical precursor of vitamin A, occurred in carrots, but we had little idea how efficiently it could be used by man.

Expert opinion in 1942, when the experiment was first planned, expected volunteers put on a diet very deficient in vitamin A to show clinical signs of deficiency in a matter of weeks, certainly in under six months. In fact after six months no signs of deficiency were apparent. Some clinical signs of deficiency appeared during the second year, and one man went for $24\frac{1}{2}$ months, without developing any recognisable symptoms.

A full account of this work has been published in Medical Research Council Special Report Series No. 264, "Vitamin A Requirement of Human Adults", so there is no need for me to repeat all the details. I am here concerned mainly with the investigation from the guinea pig's point of view. I had originally been worried lest the Committee should fail to enjoy the confidence of their experimental subjects. However, they quickly learned how to conduct the experiment, as the following extracts from the M.R.C. report show.

L. M. Rasmussen (ed.), *Human Guinea Pigs, by Kenneth Mellanby: A Reprint with Commentaries*, Philosophy and Medicine 134,
https://doi.org/10.1007/978-3-030-37697-0_16

"The volunteers were given the greatest possible share in the responsibility of running both the community and the experiment. They were encouraged to take an intelligent and active part in the work, and to regard themselves not as mere passive guinea pigs but as partners, equal in importance to the team of investigators. Many routine duties, especially domestic duties, such as cooking and cleaning, were discharged by the volunteers and everyone was expected to play his part. The most important part of the routine work in any nutritional experiment is the cooking. The ingredients to be used, as well as standard recipes and general instructions about the diet, were laid down by the investigators but the actual preparation of the meals was made the responsibility of the volunteers themselves. One of them, a professional baker and confectioner, was in charge of the general supervision of the kitchen, while the bulk of the routine cooking was in the hands of a rota of volunteers. Kitchen duties changed weekly and all the volunteers who were not employed on work outside the Institute were on the rota. This method of changing the cook introduced some variety of style and, what was more important, greatly reduced the grumbling against him. In a long drawn out nutritional experiment with a monotonous diet, the cook is liable to be the first target for grievances. By making the volunteers themselves responsible, in turn, for the preparation of the food, the meals were virtually eliminated as a source of grumbling. The prospect of taking their own turn at cooking, and the existence of a high degree of *esprit de corps*, helped the volunteers to put up cheerfully with the short-comings of the food.

"Other routine work in which the volunteers took a responsible share was laboratory work, record-keeping, and clerical work."

The most noteworthy hardship of this whole experiment was boredom. The diet, even by wartime standards, was extremely dull. For long periods no vegetables at all were allowed, and no fresh milk or any milk product except skimmed milk powder. No eggs were eaten—not even the occasional "ration" egg enjoyed by the rest of the population. A small extra ration of special vitamin-free margarine, some extra meat (fat free), jam and unfortified chocolate was included, as otherwise insufficient calories would have been available. The volunteers could not go out for a spree—they took their own provisions whenever they were away from base. They were all free to stop the experiment whenever they wished, but none did so without good reason or before the Committee wished them to stop.

In the original scabies work, there was something of a crisis of morale when, after a few months, no one had contracted the disease (p. 60). Similarly, when none of the expected symptoms of vitamin A deficiency (night blindness, skin troubles, etc.) appeared after an even longer period, morale was once more at a low ebb. The volunteers began to doubt the usefulness of their contribution. Various subjective symptoms, not usually associated with dietetic deficiency, such as fatigue and loss of memory, began to be complained of—but equally by deprived individuals and controls. This was a passing phase, but it showed that, in a prolonged experiment, strain and doubt may be confusing factors to be separated from the true results of the experiment itself.

The success of this experiment depended entirely on the reliability of the experimental subjects. There was nothing to stop them cheating in their diet—and their friends often offered them forbidden luxuries. It is really amazing that they continued so conscientiously for this unprecedented period. Incidentally, had they taken forbidden foods, the analyses of their blood and their excrements would have given the game away, but those running the trials never felt that they needed this additional safeguard (though it was useful in convincing other sceptical scientists who did not know the guinea pigs). The experiment was unique in that it continued for a far longer period than any other group trial either before or after. I should be very surprised if anything on this scale is ever repeated.

As has already been indicated, the most noteworthy finding from this experiment was that, at least in young adults, the risk of severe clinical symptoms of vitamin A deficiency developing was much less than was commonly (even by dietetic experts) believed. It did enable an accurate picture of the way these symptoms develop to be determined, and very valuable information on the best dosing régime to restore any who had suffered deficiency was obtained. From the strictly scientific point of view, a mass of unique data on the relationship between the blood levels of vitamin A and carotene, and the onset of clinical symptoms, was collected. The subjects made a considerable contribution to the understanding of the processes of adaptation of the human eye to reduced light intensity, which in extreme cases results in "night blindness".

The committee found it difficult to be dogmatic about the minimum human requirements for vitamin A. The experiments showed that about 1300 I.U. of vitamin A was the minimum protective dose for this group of people, and they recommended a daily intake of 2500 I.U. for the general population, to cover individual variations and to allow some margin of safety. They found it more difficult to resolve the question of how much carotene, the natural precursor of vitamin A, found in most vegetables as well as in some animal tissues, was required. It was known that less than 100% was transferred into vitamin A, and that under some circumstances a considerable amount might be excreted and so would not contribute to the effective diet. The general conclusion was that about three times as much carotene was needed, as compared with the required dose of preformed vitamin A.

Incidentally the experiments did not show that there was any advantage in giving large doses of vitamin A to normal individuals. No volunteer was given amounts such as are found in the liver of the polar bear, which has been found to be actually toxic, but the sort of high dose sometimes recommended by those selling vitamin concentrates were tried and found wanting. Super sensitivity of the eyes in the dark was not produced.

One interesting experiment with carrots was made before the main vitamin A experiment started. A group of volunteers lived for a few weeks on a diet containing many pounds of carrots per day. At this time the general public did believe that such a régime would enable thepi to see better in the "blackout". It did not. As the carotene level in the food increased, so did that in the faeces, and only a small fraction was absorbed. Incidentally we saw no indication of carrot eaters developing orange pigments in the skin, something which has been described in the medical literature. This may possibly occur in a small minority of the population, but man seems normally to have a simple mechanism to combat over-indulgence in this vegetable.

Today scientists often speak of the development of the "team" in scientific research, giving the idea that this is a new development. Our volunteers were the focus for the attentions of a unique collection of scientists, who came regularly to Sheffield from their bases in Cambridge, Liverpool, Oxford, Reading, and London, as well as from laboratories in Sheffield itself. The table, taken from the M.R.C. report, shows how the research team was organised.

I had been instrumental in persuading the authorities to start the vitamin A experiment, and I took some part in getting it going. Even after I was in the army, when I visited Britain, I returned to Sheffield from time to time to see the volunteers to smooth out some of the difficulties which arose in the early months before the Committee had gained the necessary experience in handling their personal problems, and sometimes to organise additional clinical trials relating to scabies. Thus the volunteers were responsible for eliminating scabies from a large mental hospital, where the disease had been endemic for years. I remember that our first knowledge of the allied invasion of France on "D-day", 1944, was over the radio when we were working in this hospital on what was very nearly my last visit to the Sorby Institute. Thereafter my contacts became more and more tenuous, and were virtually non-existent by the time the vitamin A experiment came to an end in the autumn of 1944. I can take no credit for the final experiment, on the vitamin C requirements of human adults.

Organisation of the Research Team

General Clinical Examination

J. Pemberton, M.D., M.R.C.P. H. M. Sinclair, D.M., B.Sc.

Ophthalmological Examination, including Use of the Slit Lamp

W. J. Wellwood Ferguson, M.B., Ch.B.
H. M. Sinclair, D.M., B.Sc., and other members of the Oxford Nutrition Survey.

Estimation of Blood Vitamin A and Carotenoids

S. K. Kon, D.Sc., Ph.D., F.R.I.C., Mrs. E. H. Mawson, Ph.D., M.Sc.
T. Moore, D.Sc., Ph.D., A. W. Davies, J. Tosie, Ph.D., M.Sc., and Miss A. C. Cooper.
R. A. Morton, D.Sc., Ph.D., F.R.I.C., and T. W. Goodwin, M.Sc., A.R.I.C.
H. M. Sinclair, D.M., B.Sc., and Miss C. M. Wood, B.Sc.

Blood Cell Count including Platelets, Haemoglobin Estimation

H. J. Barrie, B.M., B.Ch.

Fractional Test Meal

J. Pemberton, M.D., M.R.C.P.

Skin Biopsy

J. Pemberton, M.D., M.R.C.P., and H. J. Barrie, B.M., B.Ch.

Measurement of Capacity for Dark Adaptation

Miss C. M. Wood, B.Sc. (Adaptometer of Wald).
W. Bartley (Adaptometer of Wald and Rod Scotometry).

The late K. J. W. Craik, Ph.D., and Mrs. S. MacPherson, B.Sc. (Adaptometer of Craik).

Night Vision Scotometry

P. C. Livingston, C.B.E., A.F.C., F.R.C.S.
W. Bartley.

Estimation of Carotene in Faeces

S. K. Kon, D.Sc., Ph.D., F.R.I.C., and Mrs. E. H. Mawson. M.Sc., Ph.D.
T. Moore, D.Sc., Ph.D., A. W. Davies, J. Wierzchowski (Polish Army Medical Corps), and Miss A. C. Cooper.
R. A. Morton, D.Sc., Ph.D., F.R.I.C., and T. W. Goodwin, M.Sc., A.R.I.C.

Audiometry

J. L. Burn, M.D., B.Sc., D.P.H.

Psychological Examination

D. Russell Davis, M.D., M.R.C.P., D.P.M., and W. Bartley.

Chemical Analysis of Foods

H. A. Krebs, M.D., F.R.S.
T. F. Macrae, O.B.E., D.Sc., Ph.D., and the Royal Air Force Nutrition Team.
T. Moore, D.Sc., A. W. Davies and Miss A. C. Cooper.
R. A. Morton, D.Sc., Ph.D., F.R.I.C., and T. W. Goodwin, M.Sc., A.R.I.C.
S. K. Kon, D. Sc., Ph.D., F.R.I.C., S. Y. Thompson, Ph.D., and Miss K. M. Henry, Ph.D., A.R.I.C.

Sampling of Specimens (Blood, Faeces)

H. A. Krebs, M.D., F.R.S.

Biological Test of Vitamin A Potency of the Diet

W. A. Broom, B.Sc.

Statistical Analysis

Miss. K. H. Coward, D.Sc.
J. O. Irwin, Sc. D., D.Sc. (Results of Blood Determinations and Audiometry).

Charge of Records

H. A. Garling.

Supervision and Organization of Arrangements in Sheffield

H. A. Krebs, M.D., F.R.S.
K. Mellanby, O.B.E., D.Sc.

General Organizer and Secretary to the Team

Miss. E. M. Hume, M.A.

The disease known as scurvy has been recognised for hundreds of years. Its symptoms include weakness and lassitude, and fatigue after even slight exertion. The skin is affected, and sponginess and bleeding of the gums occur. Wounds do not heal, the joints swell and various infections are said to be difficult to control. It was once common in sailors, and among those whose diet was poor in fresh fruit and vegetables, and which contained little fresh meat. The symptoms generally cleared up if they were not too advanced when the diet improved, and the title of "limeys" given to the British derives from the practice of giving lime or lemon juice to our sailors to prevent scurvy. The name "vitamin C" was suggested some 65 years ago for the unknown substance which prevented scurvy and which occurred in fruits and other foodstuffs. It was identified as ascorbic acid, and synthesised in the laboratory, in the 1930s.

During the war clinical scurvy was reported in some individuals in Britain, and the early symptoms were quite commonly suspected in infants and young children. Several small scale human experiments had been made showing that symptoms could be produced by poor diets in a few months. As ascorbic acid is easily broken down, and as it is not stored in the body in the way in which a fat-soluble substance like vitamin A is accumulated, there was considerable concern lest deficiencies might become widespread. Thus when it became known that the vitamin A experiment was ending, and that many of the volunteers were still willing to start a new experiment, the Medical Research Council decided (or, more accurately, was persuaded) that a similar investigation relating to vitamin C should be made.

Here again it is fortunate that this investigation has been fully described in the Medical Research Council Special Report No. 280, entitled "Vitamin C Requirement of Human Adults". The following passage is taken from the Preface to that Report:

> "Estimates by different authorities of the amounts of viamin C which should ideally be present in the diets of normal adults have been widely divergent. In this country ordinary diets contain an abundance of the vitamin and signs of deficiency are very rare. During the war years, however, when foods were restricted and the limited supplies of the natural and synthetic vitamin had to be used to the best advantage, it became a matter of urgent importance to establish what daily intake was needed to maintain health, and whether any extra benefit was derived from taking larger doses.
>
> "Accordingly, in 1944, the Council arranged for a reinvestigation of the problem and asked their Accessory Food Factors Committee to organise a clinical trial on volunteers, similar to that which had just been conducted to estimate the requirements of vitamin A. The trial was directed by the Vitamin C Sub-committee and was carried out at the Sorby Research Institute, Sheffield, between October 1944 and February 1946. The Chairman, Secretary and most members of the Sub-committee travelled to Sheffield regularly in order to meet the other investigators and the volunteers. The nature of the investigation was such that fresh decisions had to be taken from time to time, but none was made until all aspects of the problem had been discussed with those engaged in the work.
>
> "Most of the twenty volunteers were conscientious objectors to military service and the Council wish to take this opportunity of thanking them for their help. Many clinical, biochemical and physiological observations were made, and the influence of vitamin C deficiency on the healing of experimentally produced wounds was studied. Scurvy developed in all those volunteers completely deprived of the vitamin but could be quickly relieved by a daily dose of 10 mg. The efficacy in the trial of this small dose provided strong additional

> evidence in support of the League of Nations figure of 30 mg as a safe daily allowance for the normal adult.
>
> "The main findings of the investigation were briefly reported in *The Lancet* in 1948, but the Council are glad to issue in their Series this detailed account containing many valuable long-term records of day-by-day observations on the individual volunteers."

Officially this experiment started in October 1944 and lasted to February 1946. Nineteen men and one woman acted as experimental subjects. Ten of these had run the full course of the vitamin A experiment. The diet in the earlier experiment had been supplemented with a daily dose of 50 mg of ascorbic acid, and it was assumed that this had prevented any risk of vitamin C deficiency. The volunteers had a rehabilitation period at least three months on an "unrestricted" (wartime) diet, and a careful clinical examination in September 1944 showed no apparent residual effects of the earlier deprivation. The other ten volunteers had not been under previous investigation, but they also showed no signs of deficiency when examined. Because of the instability of ascorbic acid, it was not thought that the diet before exposure to the experiment was a very important factor.

The volunteers were divided into three groups. One, of 3 individuals, were controls, taking 70 mg of vitamin C daily. The second group, of seven, took 10 mg per day; this was thought to be too low an intake, but was comparable with the intake of some of the population outside. The third group of ten had a diet as low in vitamin C as was possible; analyses showed that it did not contain more than 1 mg. This account of the general course of the experiment is taken from the M.R.C. Report:

> "The general management of the volunteers was along the lines described in the Report of the vitamin A trial (see p. 160 above). The volunteers could be relied upon to co-operate in every respect. They assisted in the laboratory work, keeping of records, and the general running of the Institute. They were fully aware of the nature of the experiment and were kept informed of the decisions of the Committee. However, for the first 6 months they did not know to which group they belonged, whether totally deprived of vitamin C, or supplemented with it. During the later stages, beginning in May 1945, the grouping became obvious from the appearance of the signs of scurvy."

As in the previous experiment, the food was not calculated to appeal to the gourmet. Foods known to be entirely without vitamin C are not very numerous, consisting mainly of flour, cereals, sugar and the like. Vegetables were included, but instead of the short cooking advised normally by dieticians, dehydrated potatoes, carrots, peas and beans were boiled and kept hot for hours until the ascorbic acid was destroyed (a system unfortunately still favoured in many canteens to this day). Milk had copper added and was then heated and spray dried. Other foods were similarly abused. Every attempt was subsequently made to render the meals attractive, but this proved very difficult if not impossible.

The experiment showed that the control group developed no symptoms at all. The group on the low dose, 10 mg, did not develop clinical scurvy, though their gums were affected and other sub-clinical effects were noticeable. All the deprived group developed clinical scurvy, the first symptoms appearing within four months in some cases, though they were much delayed in a few individuals. The symptoms were quite quickly relieved even by a low intake of 10 mg. A high intake acted more rapidly. The general conclusion was that the minimum intake of 30 mg previously advised by the League of Nations was reasonably high. Incidentally, several of the most seriously affected individuals asked that their cure should be delayed as long as possible, if useful observations could be made on them during this period.

One of the more unpleasant experiments was on wound healing. A 3 cm cut was made deep into the flesh of a thigh. This was allowed to heal and then the weight needed to open the wound again was measured. Deprived individuals healed less well, but no protective effect of high intakes of the vitamin were detected.

Today many people believe that a very high intake of vitamin C—many times the dose recommended from these experiments—may be advantageous, particularly in reducing the length of the period for which an individual suffers from a common cold. It is interesting to note that, on the average, all groups of volunteers in the vitamin C experiment caught the same number of colds, but that the deprived group got better rather less rapidly. No advantages of a very high intake were found, but the experiment was not planned to investigate this thoroughly.

This experiment ended in February 1946, and the volunteers started to disperse. It took some time to wind up and to complete all the analyses. The first account of the findings appeared in 1948 in *The Lancet*, and the comprehensive report five years later. However, the practical results were available to the authorities at a much earlier date.

Many of the volunteers had acted as experimental subjects for nearly six years, suffering first from scabies and then from two major dietetic deficiencies. This was a remarkable effort, and one for which the scientific community as a whole should be most grateful. I do not think that such an achievement as theirs will ever be seen again.

Chapter 15
Malaria

Malaria, a disease caused by a protozoon parasite in the blood, and transmitted by mosquitoes of the genus *Anopheles*, is the most serious disease from which the human race suffers. In peacetime the infant and child mortality of half of mankind is increased, and the working efficiency of adults is greatly decreased in the same areas. In war malaria may be the armies' most serious enemy.

Malaria disappears if the mosquito vector is eliminated. Public health measures may destroy the breeding places permanently, or the insects may be killed with insecticides. Men may also be protected by taking anti-malarial drugs which not only cure the infection but which act prophylactically and allow the victim to be bitten by infective insects without developing the symptoms of the disease. If drugs can entirely eliminate the protozoon parasite from a population, the mosquito has nothing to transmit and can therefore be ignored. All methods of malaria prevention have proved fruitful fields for research, the incidence of the disease has been reduced at least temporarily in many places, but there are still problems to study before they are completely solved. Some of these were the subject of investigations during the war in which human volunteers took an important part.

This is an important and complicated story, which I cannot deal with in detail here. An admirable account from the army point of view is contained in F. A. E. Crew's publications, particularly "The Army Medical Services during the Second World War" published by H.M. Stationery Office 1957–66, and in various papers by Sir Neil Hamilton Fairley. Here I will again deal mainly with the story as it concerned the human volunteer.

During military operations malaria prevention by mosquito control is likely to be impossible, particularly when the armies are moving rapidly forward or (as happened in South-East Asia in 1942 when we started considering the problem)

L. M. Rasmussen (ed.), *Human Guinea Pigs, by Kenneth Mellanby: A Reprint with Commentaries*, Philosophy and Medicine 134,
https://doi.org/10.1007/978-3-030-37697-0_17

backward. Also at this time efficient insecticides were scarce. All the supplies of pyrethrin, which is very effective against mosquitoes, were earmarked for mosquito control—DDT did not appear until considerably later. We had no good repellents to keep the biting insects away from their victims. Although the German drug atebrine (usually known by the trade name mepacrine) was known to be an effective curative drug there were conflicting views as to its efficiency in preventing infections from occurring, and alarming stories about its toxicity and possible effects if ingested regularly. I was a member of several Medical Research Council committees which were concerned with these problems, and I saw something of the various research programmes.

At Oxford Professor Brian Maegraith made an intensive study of mepacrine toxicity. It was being recommended that troops in malarial districts should take one tablet every day; some people said this would not give protection, others feared that the takers might be poisoned. There was some prejudice because those who took mepacrine turned yellow, and appeared to look like jaundice patients. Some troops said they all vomited regularly when taking the drug, and the usual rumours that it was given to reduce virility and potency were rife. Maegraith and his colleagues, and some 480 Oxford undergraduates, ingested enormous amounts of mepacrine under a whole variety of conditions. They also made pharmacological studies on themselves with other anti-malarial drugs. The main conclusion of the mepacrine work was that the drug could be safely taken by healthy people, and that harmful side effects were rare and could often be avoided.

At the same time work was in progress to find what degree of protection the drug conferred. Here I was personally involved. Volunteers were required, and I found that the Friends' Ambulance Units were very willing to help. I rented an office in Golden Square near to Piccadilly Circus, and this was the centre where some 300 volunteers came for regular examination. They also went to the malaria laboratory at Horton Hospital where Dr. W. D. Nicol and Mr. P. G. Shute infected them by mosquito bites and by injecting them with blood from malaria patients. They were put on various courses of mepacrine, and frequent blood samples were taken and analysed, mainly by Major James Reid of the R.A.M.C. These experiments confirmed in general that each particular level of intake of the drug gave a corresponding level in the blood, and it was on this level that protection against the parasite depended. A number of volunteers, insufficiently protected, developed clinical malaria. We were using a virulent strain of Malignant Tertian malaria, which is often fatal, particularly in those infected for the first time and not given antimalarial drugs. For scientific reasons it was often desirable to allow an infection to run for several days, but the physicians in charge of the experiments were reluctant to do this as they did not wish to endanger the volunteers any further. This fact became known to the volunteers, who protested and said that, as long as the experiments were really aimed at relieving human suffering, they wished the disease to be allowed to go on as long as the experiment required. Several said that they were prepared to sign a document exonerating the experimenters should the experiment prove fatal. They made the point that the Friends' Ambulance Units do not hesitate to act in the field during active service, and that quite a number of members have been battle casualties. Under these circumstances they considered that they should be prepared to risk their lives equally in a useful experiment.

Our malarial research in Britain was on a considerable scale, but it was rather overshadowed by that of Sir Neil Hamilton Fairley in Australia. The Australian forces had a real problem. Their troops in New Guinea, in December 1943, had some 740 cases in hospital for malaria for every thousand in their armies. The disease did more damage than the Japanese. Fairley persuaded the Australian military authorities to set up a research establishment on the Atherton tableland in Queensland. Here he had all the laboratory and scientific facilities he asked for, and an unending supply of volunteers from the army. Some people have criticised this arrangement, complaining that army volunteers usually have no power to refuse. I visited these experiments, saw something of the volunteers and have no doubt that they were not unwilling conscripts. Australian soldiers are noted for their independence of mind, and I am convinced that these took part because they wished to help to solve a really important problem.

The Australians, like our volunteers, were given various prophylactic courses of drugs, and were subjected to different degrees of infection by malarial parasites. In order to find out whether protection which seemed effective in unstressful conditions would be equally effective during hostilities, the volunteers were subjected to cold, to exhaustion and even to the effects of alcoholic excess. The main practical result showed that a recommended daily dose of mepacrine gave such a high degree of protection under all conditions that it was reasonable to court-martial any officer whose troops had any substantial amount of malaria, on the grounds that discipline was lax and the men had not been taking their drugs. As a consequence, by November 1944 the annual rate of malaria in New Guinea had fallen from 740 (in 1943) to 26 per thousand. We adopted the same measure in South-East Asia with similar results. Malaria ceased to be a serious military problem to our troops, though the Japanese, who did not use these methods, continued to suffer many malarial casualties.

These experiments showed that soldiers and conscientious objectors could give the same sort of service where this proved to be important. However, some pacifist volunteers considered these malarial experiments to be primarily of military importance, which is true, and they were therefore unwilling to participate. Most of those who had volunteered for ambulance duties did not hesitate, as they saw disease control in the troops as being similar to giving medical treatment to battle casualties. This again made one realise that the term "pacifist" covers a wide range of points of view. The conduct of the volunteers showed them to be tougher and more willing to suffer danger than most other members of the population.

As an addendum to this chapter, I may describe one little experiment in which I myself took the main part. Italy just south of Rome has a very virulent strain of malaria, and, before our troops entered this area, we wished to know whether this strain too would be susceptible to the suppressive effects of mepacrine as taken in other theatres of war. We wished therefore to import infected mosquitoes into Britain. The easiest method, we decided, would be to fly the insects straight here, which meant going over France, then occupied by the Germans. Many allied planes made this journey, usually travelling as high as they could. What we did not know was whether the mosquitoes would survive the cold and low pressure of such high altitude flights in unheated, unpressurised planes. I worked for a short time using the chamber simulating high altitudes at the Royal Aircraft Establishment at Farnborough. The low pressure, equivalent to 30,000 feet did not seem to worry mosquitoes, even if I found it a bit trying, but they soon died at the low temperatures of about −30 °F. I sat day after day under these simulated conditions in the chamber, trying out various techniques to keep the mosquitoes happy. In the end the simplest proved best. I wore an insulated flying suit several sizes too big, and cuddled a cage of *Anopheles* in my bosom. Like more experiments, this was uncomfortable and infinitely tedious rather than dangerous. And we never made use of the results, as we obtained the required strain of the parasite from a patient who had been infected in the danger area before we had been able to arrange to collect and transport the mosquitoes. But the experiment added an iota to our knowledge of the effects of low temperatures on insects!

Chapter 16
Experiments in German Concentration Camps

Hansard, for 22 January 1947, reads as follows: "Mr. Somerville Hastings (Barking: Labour) asked the Chancellor of the Duchy of Lancaster whether permission was given to Dr. Kenneth Mellanby to travel to Germany in the uniform of a British Press Correspondent to talk to German doctors now on trial at Nuremberg about their experiments on human beings, and what use will be made of his report." Mr. John B. Hynd (Athercliffe, Sheffield; Labour; Chancellor of the Duchy of Lancaster and Minister for Germany and Austria): "Dr. Mellanby travelled as an accredited representative of the *British Medical Journal*, and was given the usual facilities afforded to a Press Correspondent. The use which is made of his report is a matter for the editor of the publication concerned."

I have already mentioned something of my experiences when Parliamentary Questions ("P.Q.s") were asked about our volunteers in Sheffield. These had been so numerous, and often so trivial, that I have never since been able to treat the subject with the awe it deserves—and the flap a P.Q. produces in a government department has to be seen to be believed. Junior civil servants, who have to produce the information, are generally thrown into a state of panic by the very suspicion that a question may possibly be asked, and when it actually appears even the most essential work is dropped and masses of relevant and irrelevant information are put together for the benefit of someone rather less junior who will actually draft the reply. At a more senior level, the P.Q. is treated as a rather intellectual game, and, as Dick Taverne says in an article in the "Listener" of 22 November 1972, to such officials "the ideal answer—should be short; appear to cover the subject-matter completely; when challenged, prove accurate in every detail; give rise to a minimum number of awkward supplementary questions; and reveal precisely nothing". I am not (God forbid) a senior civil servant, but I think the answer quoted above, which

L. M. Rasmussen (ed.), *Human Guinea Pigs, by Kenneth Mellanby: A Reprint with Commentaries*, Philosophy and Medicine 134,
https://doi.org/10.1007/978-3-030-37697-0_18

was in fact written by me, indicates that I have some inkling of the principles of the game. It is short, true, and completely misleading. How it came to be asked, and answered, requires some explanation.

During the 1939–45 war we became increasingly aware that medical experiments were being done on the prisoners in German concentration camps. As mentioned below (p. 188) accounts of some of this work had actually been published in reputable scientific journals. In general, however, we had to rely on vague and horrifying rumours of torture and bestiality on an unprecedented scale.

Public reaction to these rumours was varied. Quite a number of articles and letters appeared in the British and American press in 1945 and 1946 on the subject. Some suggested that the only proper thing was to have nothing to do with such horrors, and that all records should immediately be burned without any scrutiny. The other point of view was that we should try to make use of the results. The victims were dead; if their sufferings could in any way add to medical knowledge and help others, surely this would be something that they themselves would have preferred. This was certainly my own approach.

Conditions at this time in Germany were pretty chaotic. Bomb damage had destroyed much of the cities, transport had largely broken down, supplies of food, fuel and many other substances were scarce and ill distributed. The authorities were fully stretched trying to restore a vestige of civilised conditions, as well as trying to bring the ring-leaders of the Nazi hierarchy to trial. The medical services were fully occupied and had no time to scrutinise all the papers left in the concentration camps. There was a movement, supported by many members of the allied forces as well as by some Germans, to try to wipe all signs and memories of the concentration camps from the face of the earth. There was a serious risk that all the records would be destroyed, either accidentally or as a deliberate policy. Many did in fact disappear.

No one in Britain seemed to be very concerned. I was by this time back in Britain, working in London, as Reader in Medical Entomology in the London School of Hygiene and Tropical Medicine. I was unable to get any official support from the military authorities or even from the Medical Research Council to rescue the records,[1] though everyone was reasonably sympathetic. I think that the authorities felt (perhaps with some justification) that the matter was not my responsibility. The trouble was that it was no one else's either. As a last resort I approached Dr. Hugh Clegg, editor of the *British Medical Journal*. I knew him already, as he had published much of the work on scabies described earlier in this book. He became very interested, and said he would welcome articles on the medical experiments for inclusion in his journal. He even offered to pay at a rate which would meet the cost of going to Germany. So I decided to devote my Christmas vacation to this task.

The first difficulty was that the Allied Control Commission did not readily admit unattached civilians to Germany. This was understandable, in view of the chaos prevailing. To obtain a permit would probably take months, and I would need to

[1]In fact some of the work was subsequently studied on behalf of the Medical Research Council, with conclusions similar to those expressed here.

persuade all manner of high officials that "my journey was really necessary". There seemed no chance of getting to Germany, and then of getting access to the records, for many months. At this point one helpful official came up with a brain wave. He told me that there was no difficulty in admitting newspaper correspondents, and for them there were no delays. If I could be officially commissioned as a foreign correspondent by a paper, I could be on my way in under 24 hours. So I went back to my editor, and explained the situation. The *British Medical Journal* had never before had a foreign correspondent, and probably has not had another since. However, it appears in the official lists as a "Newspaper", and so it has as much right as *The Times* to name a foreign correspondent. I had to find my old uniform, obliterate any badges of rank and sew on my sleeve a black oval with the word "Correspondent" embroidered in gold upon it. In this uniform, and with the papers quickly drawn up and handed over to me by the Control Commission, I could travel on military boats and trains in the British, French and American Zones of Germany, and could be accommodated in "Press Camps" or even in military buildings.

Germany is a large country, and I was afraid that searching for the records might be like looking for a needle in a haystack. I decided to start in Nuremberg. This was the seat of the International Military Tribunal, set up to try war criminals. The highly publicised trial of Goering, Hess and the other political and military leaders had just concluded. In the judgement, reference to medical experiments was made as follows: "The inmates [of concentration camps] were subjected to cruel experiments; victims were immersed in cold water until their body temperature was reduced to 28°C, when they died immediately. Other experiments included high-altitude experiments in pressure chambers, experiments to determine how long human beings could survive in freezing water, experiments with poison bullets, experiments with contagious diseases, and experiments dealing with sterilisation of men and women by X-rays and other methods."

A second trial, of those particularly concerned with these experiments, was about to begin before the International Military Tribunal. There were 23 defendants, twenty being themselves medically qualified, the remainder being senior administrators. I decided to attend at least the beginning of the trial, and to use this as a means of making the necessary contacts to continue my investigations.

The Military authorities transported me most efficiently to Nuremberg, where I was accommodated, with all the "real" Foreign Correspondents, in the luxurious castle formerly the home of the Fabers of pencil manufacturing fame. Views from the train and even the shortest walk through the streets of any of the German cities confirmed the impression of general devastation, squalor and privation. The Germans looked cold, ill-dressed and half starved, particularly the children. In contrast the press camp in the Faber schloss was warm, food was provided in excessive amounts and the supplies of drink seemed unlimited—and a minority of newspaper men lived up to the reputations for drunkenness which the cinema tends to perpetuate. We, as well as the troops, could collect an excessive duty-free cigarette ration, at a time when the black market value of a single cigarette was not far short of an American dollar. I found the value of cigarettes (which, as a non-smoker, I had not normally bothered to collect) when I travelled on one of the few civilian trains from

Nuremberg to Frankfurt and back. Instead of paying the fare in cash, which would have amounted to several pounds, I was advised by an American Military policeman to "give one of those bloody Krauts a packet of cigarettes". This I did, but I overpaid by tendering a pack of 200 cigarettes, for which I had paid a few shillings, instead of the expected small packet of 20. I have never regretted this piece of unintentional "generosity", and hope that the booking clerk made good use of his unexpected windfall.

I attended the trial for a good many boring days, though not to the end, for it dragged on for months and eventually most of the defendants were found guilty and given a variety of sentences. The trial was given all the dignity of the law, no doubt to bolster up its somewhat ambiguous legality. On the bench sat the four judges, from America, Britain, France and Russia, each with his national flag hung up behind him. The prisoners sat in two rows in the dock. The judges, defendants, lawyers and the public (for whom some hundred seats were provided) all had headphones, which relayed the trial simultaneously in English, French, German and Russian. The translation was generally of a high standard though certain common mistakes which arise between German and English regarding such things as the very different diseases "typhus" and "typhoid" could not be avoided. Typhus is carried by lice, and is caused by a rickettsia. Typhoid is a bacterial disease transmitted by water and food. Unfortunately the German "Typhus" means, in English "Typhoid"; in German "Fleckfieber" is correctly used for the English "typhus". When experiments on the transmission of these diseases were discussed, there was considerable confusion among the judges and lawyers. The prosecuting lawyers were mostly American, led by Brigadier-General Telford Taylor, who spoke well and with dignity, though some of his supporting attorneys lounged about and sat with their feet on the table, chewing gum and apparently trying to simulate an out-of-date courtroom movie. The German lawyers made a somewhat better impression. Their leader was Dr. Servetius, an elderly, dignified, grey-haired liberal barrister who had suffered considerably under the Nazis, but who had now undertaken what must have been an unpleasant duty which he pursued with considerable effect. Unlike the monoglot Americans, he seemed able to speak and understand all the languages used in the trial without the help of the multichannel simultaneous translation except when the witness was inaudible. His linguistic ability was apparent on one occasion, when a girl witness was ostensibly speaking in her native Polish. She spoke quietly (not surprisingly in view of the appalling horrors she described) and Dr. Servetius put on his headphones to listen to the German version. After a time he approached the bench and said that the translation of the testimony was very inaccurate. He asked permission to stand nearer so that he could hear what the witness actually said herself. He went on to suggest that the interpreter might be in difficulties, as the witness was in fact speaking in Russian and not in Polish.

At first the public part of the court was crowded and many reporters from all over the world attended. Soon, however, most of the seats were empty, and the press only turned up when some special horror was being described or when one of the more colourful defendants was being examined. The interest of the world in the Nuremberg trials soon evaporated. So did mine. I quickly realised that there was little of scientific significance to be learned there, and that my time would be better spent visiting some of the concentration camps, talking to surviving victims and examining documents. It was clear that few if any of the defendants had themselves been personally involved in experiments—their responsibility, even of the medically qualified, was as administrators. Any descriptions of experiments were at second hand. Also many who should have been in the dock were already dead, either from suicide or murder by other Germans, or they had already been executed by other allied courts. The following conclusions are therefore based on my own investigations rather than on the deliberations of the court.

I soon realised that no great medical advances were likely to appear as a result of the sufferings of the victims of these experiments. Little of the work had been properly planned, few of the investigators were competent, there was a lot of very inaccurate recording and even some deliberate falsification of results. Many of the so-called medical experiments were quite unworthy of that name, and were merely excuses for bestial cruelty. It is difficult to understand how human beings could have become so degraded as were many of those concerned.

It also soon became clear that many of the original reports of some of the experiments no longer existed, having been destroyed by the experimenters themselves when capture seemed inevitable.

There is one point about the inmates of the camps which needs to be made. We tend to have the idea that they were all innocent people who suffered because of their liberal or anti-Nazi political beliefs, or that they were Jews imprisoned purely on racial grounds. This is true of a very large number but there were others who were simply criminals or even criminal lunatics. Unfortunately these criminal elements were often those who were given some authority as they ingratiated themselves with the guards, and in fact some of the worst excesses were the responsibility of privileged prisoners. Nevertheless it was the German military authorities who were responsible for the camps, and who deliberately ordered the deaths of the majority of the inhabitants, even if the details of some of the atrocities may have been organised at a lower level.

I know of only one important and unique piece of medical research which was done in a concentration camp. This was the study on typhus immunisation by Dr. E. Ding. He really did use his victims as though they were guinea pigs. He prepared various anti-sera for the purpose of protecting people from louse-borne typhus ("Fleckfieber"). He then deliberately infected groups of people with the organisms causing the disease; some of these had been protected by his different anti-sera, some were unprotected controls. It was a well-conducted scientific experiment, and although many of those who took part died, including most of the controls, his work formed the basis not only of German, but also of British and allied anti-typhus vaccine policy. We in Britain were familiar with Ding's work, for it was published openly in a well-known German scientific journal (the *Zeitschrift für Hygiene und Infektionskrankheiten*) in 1943, and copies quickly reached Britain through neutral sources. Ding's methods were soon brought into use in many countries. In fact his vaccine was not widely used on our side, for, unlike the 1914–18 war, when many millions died of typhus, during the later part of the 1939–45 war lice were controlled with DDT in most areas where our troops operated and so typhus was prevented. On the German side DDT was not widely available until hostilities ceased, and here Ding's results were probably quite valuable. Ding was a reputable scientist, he appeared to think that the deaths caused by his experiments—probably not more than 250—were justified, remembering the five million who died of typhus during and after the previous war. He appears to have tried to use "volunteers", and some of those who took part and survived were said to have been released from prison or at least given favoured treatment. Nevertheless either his conscience preyed upon him, or he was afraid of the consequences, for he committed suicide and so avoided capture and trial. His own published account, and his laboratory notes, clearly established his guilt, even though his efforts almost certainly saved more lives than his experiments killed—and when he did the work he, like most other experts, expected typhus to be as serious as it had been 25 years earlier, so for every victim of his experiments 20,000 others might have been saved.

Another reasonably scientific piece of work was that at Dachau by Dr. Klaus Schilling on malaria. Schilling, who was over 70 years old when he did this work, was a scientist of international reputation, for some years a leading member of the League of Nations Malaria Commission. He infected over a thousand prisoners with malaria, and, according to the indictment, several hundred died of the disease. I simply do not believe this. We know that a mild strain of Benign Tertian malaria was used, and that this itself would seldom cause death even in patients in a poor state of health. Many of the victims were treated with drugs which would certainly have alleviated the symptoms, even if they did not always completely cure the infection. Klaus Schilling was said to have been a subscribing member of the Nazi party, and one who never hid his views at international scientific conferences. Nevertheless those who knew and worked with him did not think that he would be a party to serious and inhumane crimes and my investigations in 1946 do not support some of the criticisms made by some other Germans in later years. His work was done early in the war, when the British imprisoned many innocent refugees in the Isle of Man, and I do not think that Schilling's involvement with prisoners who had been locked up by his own government was necessarily a sign of depravity. The conditions at Dachau at the time were reasonably humane for a prison, quite unlike those discovered in 1944 when the allies liberated the camp. There seems little doubt that Schilling explained his work to the participants, and that they were quite genuine volunteers—if it is possible for a volunteer to exist within a prison. I personally believe that prisoners should not be used, but in America criminals in gaol did take part in tests involving malaria infection not unlike those done at Dachau by Schilling—and Americans are continuing to use such prisoner "volunteers" in substantial numbers to this day. In view of this fact, it is somewhat ironical that Schilling was brought before a United States army court early in 1946, found guilty of mass murder, and summarily hanged. Had he survived to the Nuremberg Tribunal he would probably have been acquitted or he might, at worst, have received a short sentence. Incidentally Schilling's work was, scientifically, rather disappointing. He confirmed the known properties of several antimalarial drugs, but did not make a scientific breakthrough like that accomplished in Australia by Hamilton Fairley (p. 175).

There was one Nuremberg defendant who was alleged to have been implicated in Schilling's experiments. This was Professor Rose from the Institute of Tropical Medicine in Hamburg. In his case the only offence alleged was that he had supplied the culture of *Anopheles* mosquitoes used by Schilling to transmit the malaria parasites to his victims. To the lay mind, convinced that Schilling had killed hundreds of victims, Rose was equally guilty. I knew Rose myself. He also was a patriotic German with right-wing views who preferred the effects of the Nazi party on the international standing of Germany to the acceptance of disabilities that the Treaty of Versailles had inflicted, yet I was convinced that he was not a monster. His offence was simply that of supplying cultures of insects to a scientific colleague. I knew from my own experience that this could have been done quite innocently. There is a considerable exchange of insect cultures between scientists—I even had malarial mosquitoes at the Sorby Institute in Sheffield, and I would never ask a colleague who asked for some just what he wanted them for. But even if Rose had known exactly what Schilling was doing, he would have had no reason to be concerned, as the experiments were planned in a humane and reasonably non-dangerous manner. I tactlessly said this to a German journalist at the trial, and he reported the matter to Rose's counsel. I was asked to give written evidence on the subject, which I did. Professor Rose was eventually set free. I think that it was absurd that he was ever charged.

The other experiments were, for the most part, of a very different nature. The most bizarre were those performed by a Dr. Rascher, ostensibly to improve the chances of recovery of sailors and others who had been exposed to extreme cold. In my youth, like many others, I was taught that in such cases, the patient should be warmed very slowly, and that if the skin suffered from frostbite, it should be rubbed with snow and not put into warm air or water. Today the treatment is just the reverse, and Rascher's experiments may have contributed, though clinical experience in other countries had altered the situation long before his work became widely known.

Rascher's first experiments were simple. He took prisoners more or less at random, but usually selecting healthy young males. At first there was some pretence of asking for volunteers, but this seems soon to have been dropped. The men were put into baths of ice-cold water, and held down forcibly until they went unconscious. This happened when their blood temperature fell from normal (37 °C) to about 28 °C. Various measurements—rectal temperature, pulse rate, respiration—were made during the process. The chilled victims were then treated in different ways. Some remained in the baths until they died, others were removed and warmed in various ways—air at room temperature, a hot room, plunging into a very hot bath. The best rate of recovery was almost always obtained using the hot bath—the technique of "rapid warming". Less drastic warming often meant that no recovery took place.

Once this principle was established, Rascher, always urged by Himmler (like some of those who today wish to reform research policy in Britain) to concentrate on practical results, decided to use other techniques for warming his victims. He pointed out that a hot bath may not be available if a severely chilled sailor is rescued from the ocean, so he suggested using "animal warmth". In a few cases men were rolled in blankets with warm pigs, which might have become an essential part of sea-rescue equipment. It was then decided that the most easily available source of warmth was that of the rescuing sailors themselves, so cfiilled and unchilled prisoners were rolled up together in the blankets as part of the study.

It was at this point that Rascher's macabre genius began to manifest itself. He decided that the experiment would be more interesting if the source of animal warmth was female rather than male (no doubt had one of the more notorious homosexual Nazis like Roehm, or Baldur von Schirach been in charge this change would not have taken place). Among the prisoners at Ravensbruck were a group of what were described as "gypsy prostitutes". These were recruited, whether as volunteers or not I am not sure, for the experiments. The chilled and unconscious prisoners were put into bed between two of these naked women, who were apparently instructed to use all their skills to raise the temperature. The results were dramatic. The chilled and unconscious subjects were apparently induced to copulate with their benefactors, and this process raised their body temperatures even more rapidly than the warmest bath. I did not hear whether this experiment altered the régime within the German navy.

Ultimately this story had, for its main actor, a tragic sequel. Rascher apparently found one blonde woman in the camp who he decided to reserve for his own purposes. She came to the eye of Himmler himself, and Himmler decided to give her his own personal protection. In the ensuing quarrel Rascher was shot, this was why he was not himself present in the dock at Nuremberg. However, Admiral Handloser, head of the medical services of the German armed forces, was officially charged with advocating the use of "Dr. Rascher's method" of rewarming those exposed to low temperatures. When I studied the documents, I found a curious error, which convinced me that Handloser was innocent of this particular charge. He had indeed, in an official document signed by himself, recommended the use of "rapid warming"—in German, "rascheres Wärmen". This had been translated by someone who had confused the adjective "rascher" (with a small "r") with the similarly spelt name of the notorious doctor, into "Dr. Rascher's method of rewarming". As rapid warming was also used by the British, with no knowledge of Rascher's existence, it seems equally likely that Handloser was actually relying on what was becoming common knowledge in medical circles.

Rascher's experiments, inhumane and sadistic though they were, might have contributed to essential knowledge about human reactions to low temperatures. Unfortunately we cannot rely on the accuracy of many of the results, such as the measurements of body temperatures, where unconsciousness (or even sexual activity) occurred. We know that Rascher's assistants, mostly themselves favoured prisoners, often accepted bribes from the victims, and wrote down much lower temperatures than their instruments were indicating. Clearly no good research is possible where the investigators are not themselves honest observers.

The experiments already described include most of those which might have produced valuable results; they make up only a tiny fraction of the atrocities committed ostensibly in the name of medical research. Thus Professor Hirt of Strasbourg killed many Jews, at a time before this was accepted policy, to produce a collection of skeletons to "prove" that he was dealing with an inferior race. In some camps a tattooed prisoner did not live long, as his hide was sought to make lampshades, though there was sometimes a pretence that a study of skin pathology was also made.

The Nazis had always had a policy of sterilizing those who they considered eugenically undesirable, and the presence of so many prisoners seemed a good opportunity for trying to improve and simplify the techniques used in such operations. The Polish girl whose testimony is mentioned above (p. 186) was one of the victims. Many prisoners suffered castration or had their ovaries removed by more-or-less normal techniques, but many operations were performed by unskilled, or even deliberately brutal operators, with dreadful results. X-rays were used in doses which caused immediate burns, and no doubt severe disabilities from over-exposure manifested themselves in those who survived long enough. Some were killed after X-ray treatment to study the histology of their tissues. Some of the women were given caustic injections via the vagina with the intention of destroying the ovaries or the tubes through which ovulation occurred. Simple male castrations were performed without anaesthetics, and such operations were turned into public and degrading spectacles. The incidence of post-operative sepsis was high, and prisoners who did not die were often permanently disabled. No new information of use to those who, in special circumstances, have to perform sterilizations, was obtained.

The greater number of deaths, however, were in an attempt to advance the new science of "Thanatology". This term was often used by the Nuremberg prosecutors to define "The science of producing death". When the Nazis had decided to liquidate all Jews and many other categories of prisoners, they found that this was not an easy task, as many million humans were involved. They therefore instigated experiments to discover the cheapest and easiest methods of killing their victims. Many prisoners were killed by neglect and starvation. Thousands of men, women and children were compelled to strip naked, to sort out their clothes to save the guards trouble, and then were machine-gunned to death in trenches they themselves had dug—and the earth was filled in mechanically on the dead and dying alike. But easier methods still of execution were sought.

A great many experiments on killing were made. Many substances were injected hypodermically and the results recorded. In most cases the intention was to give an intravenous injection of poison, but unskilled operators often missed the vein and greatly prolonged the sufferings of the victims. The cheapest and safest method of killing was found to be to inject phenol into the veins, and data were determined to find the smallest amount which was rapidly lethal. Victims of all ages and both sexes were chosen. Many died lingering and horrible deaths, when minimum doses and less lethal substances were used. The scientific value of such work must be minimal.

An examination of the results of these experiments reinforces the view that human experiments should only be done under circumstances which do not easily lead to abuse. It is clearly tragic that so little of value was discovered as the result of this appalling sacrifice of human lives.

Chapter 17
The Present and the Future

All the experiments so far described were done during or immediately after the 1939–45 war. In Chap. 13, the last in the original book, I discuss plans for a permanent institute where experiments on man could be made. It may be interesting to see what has happened in the 30 years since that chapter was written.

I had, in fact, a site in mind when I made the original proposal. Early in the war the Medical School of Harvard University in the United States set up a hospital near Salisbury in England. Though housed in "temporary" one-storey buildings, these had first-class laboratories and were far more luxurious and well designed than anything we were erecting ourselves. The hospital had a very high powered staff, many interested in the sort of epidemiological problems likely to arise in wartime. One of their scientists was particularly interested in scabies, and there was some collaboration with us at the Sorby Research Institute. Although the war had many years still to run, we did have some discussions about the future peace-time uses for which the hospital buildings might be suitable. It seemed to me an ideal place for human experiments, and I suggested this to the Medical Research Council. To some extent this proposal bore fruit.

Today the Harvard Hospital is famous as the centre of the Medical Research Council Common Cold Unit, and is in fact a place organised for experiments on human subjects. This work is under the direction of Dr. D. A. J. Tyrell and other scientists, and I am not the person to describe it in any detail. Over the last twenty or so years it has accommodated thousands of volunteers who have been subjected to the risk of infection with colds in studies of the viruses involved and in methods to prevent or to ameliorate infections. Pairs of volunteers are accommodated for periods of about two weeks in conditions of considerable comfort, and they are well fed. They have to remain isolated, and even on country walks must not approach

L. M. Rasmussen (ed.), *Human Guinea Pigs, by Kenneth Mellanby: A Reprint with Commentaries*, Philosophy and Medicine 134,
https://doi.org/10.1007/978-3-030-37697-0_19

anyone except their partners. Many have developed bad colds, but many have remained healthy and have enjoyed a pleasant, paid holiday. There have been students who have used the period for uninterrupted study, some have gone to try to write a book, and others go on official or unofficial honeymoons. Some have returned time after time. However, in the main this has been a place for short-time volunteers, quite unlike the long endurance tests at the Sorby Research Institute. The problem in which they are involved is fitted to this type of régime.

The Common Cold Unit seems to be the only place in Britain where long term human experiments are performed, even if the participants only suffer short exposures. Many sporadic experiments are made on man in other places. Thus drug firms test out new substances on members of their staff, and small amounts of insecticides are consumed by scientists employed by the makers to ensure that the ordinary user is not at risk. Individual scientists still expose themselves to all manner of hazards in the way they always have done, and will obviously continue to do so. But the largest number of serious experiments on man appear to be those done on convicts in the United States.

Prisoners have always seemed obvious subjects for experiment. When the death penalty was operative, scientists in many countries suggested that condemned prisoners might be given the choice between certain death by execution, or the possible chance of survival and pardon if they took part in a very dangerous experiment. Except in German concentration camps, where this policy may have operated (even if the promises of pardon were not always honoured) this system has seldom operated. But prisoners have been used for a whole series of experiments, particularly in the United States.

In all such cases, the experimenters have made real and sincere efforts to avoid the abuses rife in the German camps. They have at least gone through the motions of making sure that only genuine volunteers have been used. They have taken care not to "bribe" the volunteers, who have received no remission of sentence or extra privileges other than a feeling of usefulness and of expiation by working to help mankind. Actually many of the accounts of these experiments make curious reading. The volunteers are so public spirited, and so altruistic in their attitude, that it is difficult to understand why they have been sent to prison at all. This seems true of individuals sentenced for even the most horrible crimes. It is even suggested that being a human guinea pig may form an important part of the prisoner's rehabilitation.

As I have said several times earlier in this book, I do not myself approve of using prisoners in this way and I am glad that approval for such experiments has never been given in Britain. During the war, the Home Office refused, for instance, to consider a proposal for the War Office that anti-malarial drugs should be tried out in British gaols. The caution of the British authorities is well illustrated by an anecdote concerning my former colleague the late Professor Patrick Buxton. In the late nineteen-thirties he was studying louse populations, and developed a technique to extract all the lice from human hair. He obtained samples shorn from prisoners whose heads were shaved on admission to gaol in several African and Asian countries. He wished to obtain similar information for Britain. He wrote to the Home

Office asking to have hair samples from those admitted to our prisons. There was a long correspondence explaining the difficulties and wishing to make sure no prisoner had his rights abused. It was not until a year had passed, and much paper had changed hands, that someone in the Home Office pointed out that British prisoners had not had their heads shaved for nearly fifty years!

Although I accept that in most cases in the United States experiments in gaols have been well conducted, I do question whether any prisoner can ever really be called a "volunteer". Also I believe that the way in which things have developed in Germany should serve as a warning. It is so easy gradually to reduce one's standards, and to abuse one's position when dealing with a truly captive population. I think also that, in order to avoid abuses, most of those working with prisoners may be too humane. Useful experiments must often be unpleasant and sometimes dangerous. Such work needs to be done, but in my opinion, only genuine volunteers, intelligent enough to understand the situation, and always free to terminate their contract, should be subjected to these conditions.

Part II
Commentary Essays

Chapter 18
Human guinea-pigs: Mellanby, Pappworth and Club Regulation

Robert Baker

> The past is a foreign country; they do things differently there (Hartley 1953).
> History teaches, but it has no pupils (Gramsci 1919).

Human Guinea-Pigs

Educators naturally gravitate towards horror stories and scandals from the bad old days, to instill vigilance in their students, even as they silently underline the virtues of whatever is currently deemed acceptable. However, some of the deepest lessons that history can teach are revealed not by events so scandalous that they outraged people in the past, but by the very things that people in the past accepted as unremarkable (spittoons, or "Whites only" signs, for example) but which we today we find laughable, disgusting or otherwise unacceptable. One reason for reissuing old books is to allow readers to rediscover the foreign country that is our past and to become pupils willing to learn from it. With this in mind I ask readers to join me in exploring three books bearing the titles *Human guinea-pigs,* and *Human Guinea Pigs* that were published in Britain between 1945 and 1973, i.e., before and during the transition between an older, now obsolete understanding of the ethics of research on humans, and a newer conception, adumbrated in the 1947 Nuremberg Code, made practicable for researchers after the World Medical Association (WMA) published its 1964 Declaration of Helsinki, and later refined and consolidated in 1979–1982, when a US Congressional commission issued its *Belmont Report* (National Commission for the Protection of Human Subject of Biomedical and Behavioral Research 1978), and the Counsel of International Organizations of Medical Sciences-World Health Organization (CIOMS-WHO) began issuing its *International Guidelines for Biomedical Research Involving Human Subjects* (2016).

R. Baker (✉)
Department of Philosophy, Union College, Schenectady, New York, USA
e-mail: bakerr@union.edu

L. M. Rasmussen (ed.), *Human Guinea Pigs, by Kenneth Mellanby: A Reprint with Commentaries*, Philosophy and Medicine 134,
https://doi.org/10.1007/978-3-030-37697-0_20

The first two *Human guinea-pigs* books narrate entomologist Kenneth Mellanby's (1908–1993) experiences using conscientious objectors (COs) as "human guinea pigs" during World War II. Mellanby published the first edition of *Human guinea-pigs* immediately after the war ended in 1945.[1] Intended for a general audience, his book was well received and widely praised. A reviewer for *Health Educational Journal* lauded it as "popular science at its best" (Franklyn 1994). The BBC concurred, observing that Mellanby's book offers "an absorbing account of a valuable experience," perhaps because, as the reviewer for the *British Medical Journal* (BMJ) noted, "it is uncommon to find romance and science wedded together…[in] a very readable book…and [a] record of a considerable achievement" (Mellanby 1973, back cover). The *Times* of London chimed in, calling *Human guinea-pigs* "fascinating reading." Mellanby's narrative raised no ethical issues in the minds of reviewers who seemed unperturbed by Mellanby's metaphorical equation of human research participants with guinea pigs, i.e., animals kept in cages for the purpose of testing scientific theories. Perhaps the metaphor seemed appropriate because the humans that Mellanby called "guinea-pigs" were wartime conscientious objectors, a class of people loathed by much of the British public in 1945. From the public's perspective, and perhaps from the perspective of the COs themselves, service as a "guinea pig" offered a measure of redemption for people avoiding the battlefield as their fellow countrymen risked their lives in battles ranging around the globe. Moreover, no one seemed to find it odd that an entomologist—an expert on bugs—was directing experiments on humans. It helped that Mellanby, a research Fellow of the Royal Society, began his research on scabies, an itchy skin infestation by tiny mites. In addition to providing the subject matter for *Human guinea-pigs*, the experiments on COs yielded information about the transmission of scabies and a more effective treatment for scabies infestations, earning Mellanby a coveted OBE medal (Most Excellent Order of the British Empire).

After the first edition of *Human guinea-pigs* went out of print, Mellanby was shocked to learn that one "M. H. Pappworth" had "published in 1967 [a book] with exactly the same title" (Mellanby 1973, 16). Mellanby condemned Pappworth's use of his title as "an act of plagiarism on [Pappworth's] part,"—although, strictly speaking, the titles were not "exactly the same." Pappworth had capitalized all three words in his title, "*Human Guinea Pigs*," whereas Mellanby titled his book "*Human guinea-pigs*," with a hyphen between "guinea" and "pigs;" moreover, Pappworth (1967a, b)[2] added a subtitle, "Experimentation on Man," whereas Mellanby's book had no subtitle. Mellanby (1973) also complained that he could "find no mention of most important human experiments" (16) in Pappworth's book, including, of course, his own research on scabies. Had Mellanby actually read Pappworth's book he might not have wished to have his experiments mentioned, since Pappworth's focus was *unethical experiments*, performed without the informed consent of those sub-

[1] All references are to the 1973 edition unless otherwise specified.

[2] All citations to the Boston edition but also published (London: Routledge & Kegan Paul Limited, 1967).

jected to them. In any event Pappworth's "purloined" use of his title seemed to goad Mellanby into publishing an updated edition of *Human guinea-pigs* in 1973 (Mellanby 1973, 16). That edition—which Lisa Rasmussen is reissuing with commentaries in the Philosophy and Medicine book series—includes Mellanby's reflections on the Nuremberg Trial, which he had attended as a correspondent for the BMJ.

It is perhaps not surprising that, in Britain of the 1940s, a gentleman like Mellanby did not deign to read the substance of a book by a nonentity like Pappworth: a nobody who had neither an important title nor a major appointment, and who, by virtue of his birthplace, family name, religion, and education was self-evidently not anyone who counted. In contrast, like his eminent uncle, Sir Edward Mellanby (1884–1955), discoverer of vitamin D, Kenneth Mellanby was an alumnus of Barnard Castle School (endowed 1221, founded 1883) and Cambridge University (founded 1441) and was self-evidently a gentleman by the all-important measures of family birthright, religion, and education. Pappworth was a mere Jewish upstart who earned his Bachelors' degrees in medicine and surgery from a newly founded red brick school, the University of Liverpool—an institution whose degree granting authority was first authorized in the twentieth century (1903). When Pappworth moved to London in 1939, these "deficiencies" were made evident to him as he found himself barred from appointments to London's teaching hospitals because, as he was informed, such positions were reserved for gentlemen and "no Jew could ever be a gentleman."[3]

Both Mellanby and Pappworth served in the Royal Army Medical Corps (RAMC) during the Second World War. Mellanby performed scientific research in Asia; Pappworth was a military physician in Africa, Italy, Greece, and, lastly, in India, where he ran a British general hospital. After retiring with the rank of lieutenant colonel, Pappworth returned to London where he was again refused appointments at teaching hospitals and so was forced to go into private practice and to offer tutorials preparing medical students to take qualifying exams. From his outsider's perch, Pappworth had the perspective to see what gentlemanly insiders were blind to: that British physicians were unethically exploiting their patients, using them as unconsenting "human guinea pigs." Pappworth attempted to alert the British public to his findings in an article, *"Human Guinea Pigs: A Warning," published Twentieth Century Magazine* (Pappworth 1962). When this and other efforts failed to gain the attention of the British National Health Service (NHS), Pappworth published the 1967 book whose title vexed Mellanby, *Human Guinea Pigs: Experimentation on Man.*

[3] Pappworth reportedly kept this letter on his desk for the rest of his life (Booth 1994). Pappworth's daughter tells a variant of this story claiming that in an interview earlier in his career, in Liverpool, the interviewer said to Pappworth, "No Jew can ever be a gentleman and I do not want one working for me" (Seldon 2017, p. 56). It is not clear whether this is two versions of one incident or accounts of two separate incidents. *What is clear is that Pappworth, an observant Jew, confronted anti-Semitism throughout his life. One documentable fact is that although Pappworth passed the exam for admission to the Royal College of Physicians (RCP, founded 1518) in 1936, he was not elected a fellow of the college—an indicator of professional achievement normally awarded about decade or so after passing the exam—until months before his death in 1994, an unheard-of 57-year delay.*

Pre-Nuremberg British Research Ethics

Neither the British nor the Americans had an official code addressing the ethics of research on human subjects prior to the Nuremberg War Crimes Trials. Yet even though no official body endorsed a formal code of research ethics, there were relevant precedents. These were created, in part, because in the English-speaking cultural sphere "the law" encompasses more than statutes enacted by legislators; it also includes "common law," i.e., practices and traditions that law courts recognize as legally enforceable precedents. An early effort at codifying common law precedents with respect to therapeutic experiments on humans occurred in 1665, immediately after the British conquered a Dutch colony that they renamed "New York." Seeking to develop a preliminary codification of British law for the new colony, the British governor convened a panel of experts whose preliminary codification of laws—later known as "The Duke of York's Laws"—included a law requiring colonial physicians, surgeons, and midwives to follow "known and approved Rules of Art" in caring for their patients. These healthcare practitioners could, however, "Acte Contrary" to these rules, i.e., they could use innovative or experimental therapies on their patients if, and only if, they also, (i) consulted with those "Skillfull in the Art" and (ii) obtained the patient's consent to trying an experimental treatment.[4] Thus as early as the seventeenth century, British common law was understood to require medical professionals to engage in peer consultation and obtain a patient's consent before subjecting a patient to an innovative or experimental treatment. In the 1767 case of *Slater v Baker and Stapleton*, for example, a surgeon was fined for using an experimental treatment on a patient in a manner contrary to the standard practice of the profession without the patient's knowledge or consent. These two safeguards—consultation with other healthcare providers and consent from a patient—would become important elements in Anglo-American expectations for ethical and lawful experimentation on humans.

Lady Mary Wortley Montagu (1689–1762), wife of the British consul in Constantinople, set another eighteenth-century precedent. During her residence in Constantinople Lady Mary, a once beautiful woman whose face had been ravaged by small pox, became intrigued by the Turkish practice of inoculating their children with pus from the sores of small pox victims to immunize them against contracting small pox (a process later known as variolation). In 1717 she wrote to a friend that "you may believe I am well satisfied of the safety of this experiment, since I intend to try it on my dear little son. I am patriot enough to take the pains to bring this useful invention into fashion in England, and I should not fail to write to some of our Doctors very particularly about it" (Lady Mary Montagu 1718). Since her son was

[4] To quote this provision as originally stated: no one "Employed about the Bed of Men, women, or Children... for the preservation of Life or health as Chirurgions, [i.e., surgeons,] Midwives, [or] Physicians, [may] presume to Exercise, or put forth any Acte Contrary to the known approved Rules of Art... without the advice and Counsell of such as are Skillful in the same Art ... and Consent of the patient or patients if they be Mentis Compotes... which law... is...intended... to inhibit and restrain presumptuous arrogance" (Duke of York's Laws).

unharmed by inoculation with small pox pus, on returning to England Lady Mary engaged a Scottish physician, Charles Maitland (1668–1748), to inoculate her 4-year-old daughter in the same manner. She also arranged to have the inoculation witnessed by three surgeons from the Royal College of Physicians (Maitland 1722). After Lady Mary's daughter survived unharmed, by Royal command on August 9th· 1721, Maitland demonstrated the safety and efficacy of small pox inoculation by testing it on six prisoners facing the death penalty. After they survived unscathed, Maitland tested their immunity by having some of them exposed to smallpox (Glynn and Glynn 2004). They proved immune (and their sentences were commuted to exile). The safety and efficacy of inoculation having been established, the British Royal Family was soon inoculated, and the practice of inoculation became widespread—even though some other people inoculated with small pox pus contracted the disease.

This "Royal Experiment" was widely publicized and set several precedents. Importantly, unlike the therapeutic experiments addressed in the Duke of York's law, this experiment was not aimed at healing the sick; instead it put healthy people's health at risk in order to expand medical knowledge. More to the point of this essay, it also set the precedent of using prisoners as "human guinea pigs." Furthermore, as is evident from her letter, Lady Mary believed that her willingness to try an experimental preventative on her "dear children" legitimized testing it on others. The underlying idea, which the Royal Family and the Royal Academy of Physicians seemed to accept, was that by virtue of the Biblical Golden Rule (Matthew 7:12), as translated in the authorized 1611 King James version of the Bible—"all things whatsoever ye would that men should do to you: do ye even so to them"— it seemed to follow that if you would allow others to experiment on you, or those dear to you, it should be permissible for you to experiment on others. Moreover, since from the seventeenth through the nineteenth centuries, English household heads could "consent" on behalf of their extended families—not only one's children, but also one's servants and slaves—a household head's willingness to risk some "dear" family members as experimental subjects legitimized, in the experimenters' minds and in the minds of the community, similar experiments on various household members and dependents. Members of the medical community, who often experimented on themselves, freely experimented on their wives, their children, their students, their servants, and slaves and others "dear" to them. And, by virtue of their willingness to sacrifice themselves and dear ones, they felt free to offer experimental therapies to their patients and to the public. Thus, eighteenth-century colonial physician Zabdiel Boylston (1676–1766) established his right to offer experimental inoculations for small pox (variolation) to the public on the grounds that he had inoculated his own son and his household slaves.

During the eighteenth century, as the population began to shift from farms to increasingly industrialized cities, new or newly expanded institutions—charity hospitals/infirmaries, orphanages, foundling homes—were formed to meet the emerging needs of the urban poor. Mindful of the fact that in these institutions, trustees, guardians, and physicians have significant and often total control over inmates' daily lives, moral philosophers and theologians grew concerned for inmates' welfare: particularly physicians' exploitation of inmates as "human guinea pigs." Among those writing on the subject was a philosophically trained Edinburgh medical school professor, John Gregory (1724–1773). Turning the Golden Rule around, Gregory invoked it, not to legitimize experiments on human subjects, but to restrain his medical students' temptation to engage in unbridled experimentation on hapless hospital/infirmary patients. He urged students to be skeptical of "dangerous Experiments, such as the inoculating people with matter taken from one who has the very bad kind of confluent Small pox … I would ask that man if he would have done so to his Child, if he would not do it to his own Child, why should he indanger the lives of other people?" (McCullough 1998).

Similar concerns were voiced in England, where the Evangelical minister, moral philosopher, and abolitionist, Thomas Gisborne (1758–1846) condemned "rash, hastily adopted … ignorant … careless … obstinate experimenters" who were all too willing to countenance the "death of an obscure, indigent, and quickly forgotten individual," in order to test some new treatment (Gisborne 1794). Sharing his concerns was another English abolitionist and moral reformer, the Unitarian dissenter and Manchester physician, Thomas Percival (1740–1804): the man who introduced the expression "medical ethics" into the English language in his eponymous book *Medical Ethics* (1803). Alarmed by the prevalence of unbridled experimentation on the vulnerable populations in hospitals and other total institutions, Percival penned the following passage in *Medical Ethics*:

> Whenever cases occur… in which the ordinary modes of practice have been attempted without success, it is for the public good…that *new remedies* and *new methods of chirurgical* [i.e., surgical] *treatment* should be devised. But in the accommodation of this salutary purpose, the gentlemen of the [hospital] faculty should be scrupulously and conscientiously governed by sound reason, just analogy, or well authenticated facts. And no such trials should be instituted without a previous consultation of the physicians or surgeons, according to the nature of the case (Percival 1985).

Some 166 years later, American research ethics pioneer Henry Beecher (1904–1976) cited this passage as presciently recognizing that an ethical "innovator must, prior to [conducting a] study consult with his peers" (Beecher 1970). Yet, although from the nineteenth century onwards Beecher and other Americans would occasionally refer to the writings of Gregory or Percival, in Britain their works gathered dust on the back shelves of libraries and had no impact on British medical morality or medical practice. The sole nineteenth-century British disciple of Gregory and Percival was an Irish Roman Catholic University of London professor, Michael Ryan (d. 1840). A perennial outsider whose views had little influence on the London medical community of his day, Ryan was the first person to introduce Percival's ideas into British medical school lectures and the first to claim the title, "Professor of Medical Ethics." In 1831 he published lectures extolling the virtue of practicing medicine cautiously. This virtue, Ryan observed, was inconsistent with researchers' claims that their own willingness to serve as experimental subjects gave them the right to experiment on other people.

> The duty of caution in practice means 'care not to expose the sick to any unnecessary danger.' The best rule of conduct, on this important point, is the simple and comprehensive,

religious, and moral precept, 'Do unto others as you would that they should do unto you.' Whatever the practitioner does or advises to be done for the good of his patient, and what he would do in his own case, or in the case of those who are dearest to him—if he or they were in the same situation—is not only justifiable on his part, but it is his indispensable duty to do.... But if he administers a dangerous medicine, merely to gratify his own curiosity, or zeal for science, to ascertain the comparative advantage or disadvantage of some new remedy...he is guilty of a breach of ethics...and a great breach of trust towards his patient; and if the patient dies he might be severely punished.

Medical men have tried the most dangerous experiments on themselves, from their zeal for science, and even sacrificed their lives; but patients, in general, have no such zeal for science—no ambitions for such a crown of martyrdom—and generally employ and pay their medical attendants for the very opposite purposes. It must be admitted, that men who would try experiments on themselves, would be very apt to try experiments on their patients. It is a melancholy truth, but cannot be denied. The profession has always reprobated such conduct; and the medical phrase of reproach and contempt for it is "*corio humano ludere*," to play with human hide, abundantly testifies in what abomination it has been held by the faculty. It is unnecessary to dwell on this point in this age because all experiments are made upon inferior animals (Ryan 1831a).[5]

Ryan's reflections on the ethics of experimentation received a scathing review in a new medical journal, *The Lancet* (founded 1823 and still publishing). Skewering Ryan's statements on "the propriety or impropriety of making clinical experiments with new remedies," the reviewer posed the following question to readers: "Will it be believed that Dr. Ryan can be so ignorant, as to be unaware that to the experiments he repudiates we are indebted for the discovery of ... the therapeutic effects of all our remedial agents? As to the performance of *therapeutic* experiments on *inferior* animals – such a thing is scarcely heard of" (Anonymous 1831a, b). Ryan wrote in response that, "I am censured ... for having stated the rule laid down by the profession... that dangerous experiments should not be made on the sick without their consent." "Perhaps in his zeal for science" Ryan suggested that the reviewer "would allow a few experiments to be made on himself! ...Or would he prefer the application of these things on the poor?" (Ryan 1831b, 224). The *Lancet* reviewer replied by "appeal[ing] to our practical readers. How would they investigate the *curative* powers on any unknown substance? The obvious answer is, by its administration to persons in a state of disease, in minute and gradually increasing doses" (Anonymous 1831a, b, 221–222; 225).

Between them, the *Lancet* reviewer and Ryan paint a fairly accurate description of the ethics of medical experimentation in nineteenth century and early twentieth century Britain. Almost all experimentation was small scale and presumptively "therapeutic" in nature, i.e., an innovation to find more effective ways of treating patients' ailments. Self-experimentation was commonplace but, for the most part, those subjected to experiments were poor people receiving treatment in charity hospitals or other total institutions. Moreover, the process of applying incremental dosages described by the *Lancet* reviewer was a standard approach to trying new therapies. Non-therapeutic experiments were typically conducted either through

[5]A more accessible edition of Ryan's lectures is *Michael Ryan's Writings on Medical Ethics*, eds. Howard Brody, Zahara Meghani, and Kimberly Greenwald (2009).

self-experimentation or through experiments on children and people of lower status, especially those in charity hospitals, orphanages, and similar institutions. Edward Jenner (1749–1823), for example, established that vaccination with cowpox pus was safer way to immunize people against small pox than variolation (with small pox pus) through a series of experiments conducted on his own child and other people's children.

As for testing new drugs on animals, as the *Lancet* reviewer claimed, this was not standard practice in Britain in the 1830s. At that time medical schools typically used animals to teach anatomy; any physiological observations instructors might offer would be presented as anecdotal appendages to the more important anatomical lesson being taught. In one of those quirks of history, physiology first emerged as a separate discipline in Britain as a byproduct of an antivivisectionist campaign to ban painful experiments and demonstrations on living animals. An 1875 Royal Commission on Vivisection formed in response to this campaign recommended an oversight board to license anyone conducting physiological experiments on animals. Since physiologists had licenses, and anatomists did not, physiology was inadvertently established as a separate medical discipline. The number of licensed physiologists rapidly expanded from three to nineteen and these nineteen, in turn, founded a dining club, The Physiology Society. Within three years that scientific society began publishing *The Journal of Physiology* (1878–present). Ultimate validation for using living animals as models for physiological processes came in 1901 when German physiologist Emil von Behring (1854–1917) won the very first Nobel Prize for experiments on guinea pigs—thereby providing the metaphor for Mellanby's title, "Human guinea-pigs." Ironically, the British Cruelty to Animals act of 1876 gave non-human British guinea pigs official protection against abuse; their human counterparts had no such protections. Hence, Mellanby would have had more constraints on experiments on actual guinea pigs than he had in experimenting on his "human guinea-pigs."

Mellanby's Research Ethics

One important reason for reissuing Mellanby's book is that it is *not* a horror story; it is a pre-Nuremberg narrative of award-winning experimentation on human subjects that was publicly accepted and commended. His narrative exemplifies basic ethical precepts for experimentation on human subjects as they were understood before Nuremberg and prior to the WMA's Declaration of Helsinki. It reveals a mid-twentieth-century British researcher still following Lady Mary's (eighteenth-century) practice of legitimizing his experiments through Golden Rule self-experimentation, i.e., by subjecting himself to the same regimens he asked of his experimental subjects. It shows further that he practiced Percivalean peer review by submitting his research proposals to the Medical Research Council (founded 1913) for approval and funding. Moreover, as Mellanby emphasizes, even though his subjects were COs whose choices for alternative non-military service were limited, they were still *volunteers*—he uses the word "volunteer" in the book's dedication, and over 270 times in his 1945 text. Moreover, Mellanby paid his volunteers wages, and provided them with medical care, room, and board. Furthermore, he welcomed transparency, responded to Parliamentary inquiries, and invited visits from the press.

Not surprisingly, therefore, in the first edition of *Human guinea-pigs* Mellanby treats his experiments on COs as a model of ethical experimentation, proudly proclaiming that he had "usually gone through with any experiment once before asking a volunteer to follow suit, but after that I have been the observer and not the guinea pig" (Mellanby 1973, 151). Mellanby also treats researcher self-experimentation as a hallmark of morality. Thus he observes that he willingly "organised a supply of volunteers from the Friends Ambulance Unit to serve as subjects for malaria work under the direction of Professor Brian Maegraith," because Maegraith "had made trials of the toxicity of anti-malarial drugs *on himself and his colleagues*, and on some 480 Oxford undergraduates" (Mellanby 1973, 151). The clear implication is that since Professor Maegraith was willing to treat himself and his colleagues (those dear to or dependent upon him) as experimental subjects, it was morally permissible for him to conduct similar experiments on 480 Oxford students—as well as the COs that Mellanby recruited for Maegraith from the Friends Ambulance Unit.

However, Mellanby also recognizes the limitations of self-experimentation. As he notes, "Very often, as in the early work on malaria and yellow fever, the scientific workers have been their own victims" (Mellanby 1973, 151). Moreover, as a matter of methodology, "The man in charge of an experiment should be making objective observations, which he cannot do if he is also the subject of the experiment" (Mellanby 1973, 151). Finally, Mellanby (1973) remarks, researchers themselves are, "probably too valuable to waste as a mere guinea pig" (151).

In these comments Mellanby seems to reference well-known yellow fever experiments conducted in Cuba in 1900. US Army Surgeon General George M. Sternberg (1838–1915) commissioned these experiments in the aftermath of the Spanish-American War of 1898 to address a yellow fever epidemic that was decimating the American troops occupying Cuba. The project was directed by two Johns Hopkins' alumni: Major James Carroll (1854–1907), and Major Walter Reed (1851–1902), who worked as a team with surgeon Jesse Lazear (1866–1900) and Cuban bacteriologist, Aristides Agramonte y Simoni (1838–1931). As a moral preliminary to experimenting on other people, the researchers decided that they would follow the Golden Rule and engage in self-experimentation as "The best atonement we could offer for endangering the lives of others… and it was agreed … that we would all be bitten by contaminated mosquitoes" (Carroll 1906). Carroll and Lazear were the first to be bitten by the mosquitos. Both were stricken with yellow fever. Carroll survived. Lazear died.

Recognizing that, to quote Mellanby (1973), research team members were "too valuable to waste as a mere guinea pig," (151) Surgeon General Sternberg immediately forbade any further self-experimentation. This forced Reed and his colleagues to find a way to conduct life-threatening non-therapeutic experiments on people who had never previously been exposed to the disease—and thus had never developed immunity to it. It soon became apparent that the only suitable volunteers were American soldiers and Spanish immigrants who had recently arrived in Cuba. Seeking permission to recruit volunteers from the Spanish immigrant community, Reed met with the Spanish Consul to Cuba. As he later reported in a letter to Sternberg "the Spanish consul sees no reason why the observations should not be made on anyone who gives his full written consent, after the danger has been explained to him" (Reed 1900). Sternberg agreed, stipulating that no new experiment "be made upon any individual without his full knowledge and consent" (Sternberg 1900).

Subsequently both the US soldiers and the Spanish immigrants who volunteered for the experiments signed forms in which they consented

> to experiments… understand[ing] perfectly well that in case of the development of yellow fever in him, that [they] endanger [their] life to a certain extent but it being entirely impossible for [them] to avoid the infection during [their] stay in this island, [they] prefer to take the chance of contracting it intentionally in the belief that [they] will receive … the greatest care and the most skillful medical service (Benigno 1900).[6]

As compensation volunteers would receive $100 in American gold, and, if they contracted yellow fever, a further sum of $100 in American gold. Should they die, their heirs would receive two hundred American dollars in American gold coins. These large sums of money were offered in ""atonement … for endangering the lives of others" (Carroll 1906). When Reed's team established the urban dwelling *Aedes aegypti* mosquito as the vector for yellow fever, Reed and his team—and those who volunteered to serve as its experimental subjects—were heralded as heroes (Lederer 2008).

After the accolades of the popular press became yesterday's news, the Reed-Carroll yellow fever experiments were etched into the memories of generations of American and British medical students through successive editions of an influential textbook. We tend to take textbooks for granted. Yet a moment's reflection will reveal that they serve as each generation's introduction to whatever is accepted as knowledge or ethical standards by the previous generation: communicating the written and unwritten standards of what is known, what is accepted, and what is unacceptable. In this case, Edelson (1994) notes that the pivotal textbook was *The Principles and Practice of Medicine*, by Baronet Sir William Osler (1849–1919), co-founder of the Johns Hopkins School of Medicine (founded 1893) and Regis Chair of Medicine at Oxford University (1905–1919). Osler cited Reed's practices as a model for non-therapeutic experimentation on human subjects. As he puts this point, "risk to the individual may be taken [only with the subject's] consent and full knowledge of the circumstances" (Osler 1901). At international conferences Osler (1907) lauded "the yellow fever experiments in Cuba under the direction of Reed and Carroll," as excellent examples of ethical experimentation on human subjects.

---

[6] Spanish language version available at the same archive. None of the contracts signed by soldiers have been preserved.

"We have no right to use patients entrusted to our care for the purposes of experimentation," Osler proclaimed,

> unless direct benefit to the individual is likely to follow. Once this limit is transgressed, the sacred cord which binds physician and patient snaps instantly. Risk to the individual [in non-therapeutic experiments] may be taken with his consent and full knowledge of the circumstances [as was done in] the yellow fever experiments in Cuba under the direction of Reed and Carroll (Osler 1907, 8).

Osler effectively established the Reed-Carroll experiments as the gold standard for ethical non-therapeutic experiments and Mellanby, who refers to his experimental subjects as "volunteers" almost three hundred times in the second edition of this book, initially aspired to adhere to the Osler' Reed-Carroll model. Thus, Mellanby (1973) provided his volunteers with room, board, and medical treatments, paid them a stipend, and initially envisioned a written contract (49); i.e., he

> ...imagin[ed] that it would be a necessary safeguard for each man to sign some sort of detailed contract, setting forth his duties and the risks he was taking, in order to cover me in an emergency and to have something to which to hold a recalcitrant subject if he proved non-co-operative. When I made some investigations into the legal side of the question I gathered that certain risks were being taken by me in using human volunteers. No matter what he had signed, if any subject started getting nasty he could place us in an awkward position, but if our volunteers appeared to be co-operative and wrote and said they wished to take part in this sort of experiment we would probably be as well covered as by having all manner of complicated (and costly) legal documents signed and sealed and so forth. I know now that I have been lucky with my volunteers, but I have never felt that it would have been an advantage to have formalised our relations in this legalistic way (Mellanby 1973, 51–52).

Thus, ignoring Osler's recommendations and Reed-Carroll's practices, Mellanby decided that the sole reason for having volunteers sign a written contract was that such a written contract would provide legal protection for *him* as a researcher: not to ask *them* formal permission for putting *them* at risk of harm, or to inform *them* of these risks. At one point, Mellanby (1973) excuses his lack of a written contract for the initial scabies study on grounds of his own ignorance and the experiment's presumed harmlessness, i.e., he "did not know quite what I was letting my volunteers in for when I originally asked them to submit to infection with scabies. I knew that the disease could be very unpleasant and had heard patients saying that the intolerable itching had prevented them from sleeping for weeks, but it seemed unlikely that any permanent harm would be done to any of the volunteers" (51). In the end Mellanby concluded that, having been given official permission to perform his experiments, and having volunteered to serve as a subject himself, any remaining moral issues would have "been dealt with...[by] an approved code of practice...laid down by the Medical Research Council [MRC (founded 1913–1919)] and other authorities" (Mellanby 1973, 16).

Yet in the 1940s neither the MRC nor any other official group or medical association in Britain (or in America) had such code of ethical practice. Absences speak louder than words: The striking paucity of such words as "ethics,"[7] "moral"[8] or "consent," in Mellanby's text speaks volumes. Not a single line in the text of *Human guinea-pigs* speaks to any concerns about the ethical or moral rights of Mellanby's human guinea pigs. The vacuum of verbiage underlines Mellanby's apparent lack of concern. To reiterate: when Mellanby contemplated asking his human subjects to sign a contract, his concern was not to delineate his responsibilities to those whose health he was putting in jeopardy, it was rather to spell out to the experimental subjects the "duties and the risks he was taking, in order to cover *me* [i.e., Mellanby] in an emergency and to have something to which to hold a recalcitrant subject if he proved non-co-operative" (Mellanby 1973, 51, emphasis added). In the end Mellanby saw no need for "costly" consent forms and did not use any.

Mellanby claims, however, to have given his volunteers some idea of the experiments for which they were volunteering. Thus, he reports, that when asked "how I 'managed' the volunteers and made them carry out various duties" (Mellanby 1973, 52), he replied, "One thing only was essential in the managing of the volunteers in these sort of experiments, and that was to take the trouble to explain in detail just what was the purpose of anything we asked them to do. I always made a practice of holding periodical meetings which they all attended, and where I gave a progress report and answered the questions which they had asked" (Mellanby 1973, 54). However, "on occasion," Mellanby (1973) admits, "it was necessary to conceal some information, particularly when investigating such subjects as individual reactions to the presence of the causative mites," i.e., he did not inform the volunteers in the scabies experiments of the physiological consequences of being infested with mites (54).

When British historian Jenny Hazelgrove interviewed British physicians about "consenting" patients during the 1940s and 1950s, she found the same broad non-specific conception of "volunteering" was commonplace. One researcher recalls the words he used to ask patients to volunteer for experimental hepatic catheterization: "I used to go to them and say, 'Professor [S] wants to test something.'" They would reply, "'Oh that's all right doctor…you don't need to tell me all that—anything Professor S says.' And that was the attitude. Doctors were heroes at that time" (Hazelgrove 2004, 186). Hazelgrove notes that medical anthropologist Sharon Kaufman, who had explored these same issues in America during that era, found that "patients agreed to participate in research on the basis of trust, not on the basis of informed consent. Patients trusted their physicians not to harm them, to do

[7] Mellanby uses the term "ethical" or "ethics" twice in the 1973 edition on *Human-guinea pigs*: once characterizing the scabies and other experiments he supervised as "performed in a reasonably ethical manner," in contrast to those conducted in Nazi concentration camps (16), and once to characterize CO's adherence to "Christian beliefs and ethics" (135). The latter use is the sole occurrence of the word "ethics" in the 1945 edition.

[8] The word "moral" or morality is used twice in the 1973 edition of *Human guinea- pigs*: once to speak facetiously of "the morality of the young" (62) and later in reference to an "increase in moral laxity in the middle thirties" (111). Both uses appear in the 1945 edition; in neither case is the word "moral" used to refer to moral issues raised by risking human health or lives in non-therapeutic experimentation on human subjects.

something positive for them" (Kaufman 1997). This broad notion of "volunteering" could be stretched quite thin as is evident from an interview that Hazelgrove conducted with a physician who reported that as a student she witnessed

> a professor of medicine who used to ask whether patients before they went under the anesthesia 'would mind while you're asleep if I took a sample of your blood for research purposes?' (They would say) 'Of course doctor, only too glad to help.' But what he didn't say was that the blood was going to come from inside of the heart, and he was going to push a cardiac catheter up the vein into the heart, and this had a certain mortality rate; he didn't mention this (Hazelgrove 2004, 187–188).

Such experiments would have failed Gregory's version of the Golden Rule test, i.e., in all probability the researcher would not have agreed to have his own child serve as a human subject in his experiment. Yet as Ryan cogently argued in his long-forgotten lectures, even had a researcher agreed to let himself or his child serve as a subject, his willingness to do so does not entail that others would wish to subject themselves to the same experiment. People willingly restrict their own diets to become stone-age carnivores or vegans; they train for marathons, they practice Tai Chi or yoga meditation to cultivate their inner selves, and they blow themselves up in the name of some moral, religious, or political cause. Yet irrespective of whether you are a carnivore or a vegan, a scientist or a suicide bomber, your willingness to risk or to sacrifice your life for some purpose or cause does not entitle you to risk the lives of others to advance that cause. Conversely—Gregory was wrong—one's unwillingness to impose a lifestyle or medical experiment on one's self does not delegitimize someone else's engagement with that lifestyle or experimenting on others—at least not insofar as the experimental subjects are fully informed and willingly consent to serve as "human guinea pigs."

Mellanby's Response to the Nuremberg Trials

Moral standards began to change at, and in the aftermath of, the Nuremberg trials. As a researcher who supervised some of the vitamin experiments at Mellanby's Institute later acknowledged, "the strict requirements of the [WMA's] Declaration of Helsinki on medical research involving human subjects, [make] it… unlikely that a similar opportunity for human experimentation under the conditions and on the scale of experiments [at Mellanby's Institute] will be available in the foreseeable future" (Pemberton 2006, 556). The transition from an informal, loosely regulated, researcher morality predicated on trust that researchers would intuitively do the right thing (illustrated in Mellanby's book), to a newer ethics formalized through regulation and monitoring by review committees began as a delayed response to the1946–1947 Nuremberg Doctor's Trial, (*US vs. Karl Brandt,* et al.1946) This was the second in a series of trials designed to hold Germans—in this case twenty-three German physicians and administrators—accountable for crimes against humanity, war crimes, and membership in a criminal organization (the *SS*, or *Schutzstaffel*, i.e., the Nazi group that had primary responsibility for implementing the Holocaust).

Yet, horrible as the revelations at the Nuremberg Doctor's Trial were, the trial would have amounted to little more than a grisly footnote to an atrocious chapter in Western history had not it slowly dawned on Americans, Britons, and citizens of other countries that eminently respectable, well-published researchers could act towards their human subjects in a manner uncomfortably similar to the way that Nazi researchers acted towards concentration camp inmates.

During the Nazi era some of the experiments that German researchers conducted were predicted on Nazi racial and eugenic theories. These include experiments with mass sterilization techniques at the Auschwitz and Ravensbrueck concentration camps, as well as "thanatological" experiments designed to discover means of efficiently eliminating people with mental and physical disabilities, so-called "useless eaters," that were conducted at the same camps. Additionally, in the name of racial science, 112 Jews at Auschwitz were killed for a skeleton collection that was part of an anatomical project at the State University of Strasbourg designed to display Jewish racial inferiority. The Tribunal had no difficulty unequivocally condemning doctors and administrators involved in these eugenic, racial, and thanatological projects.

Experiments on human subjects directed at biomedical issues related to the war itself were not so easily condemned. Armies confronting each other on the same battlefields encountered many of the same biomedical issues. Quite naturally, therefore, their scientists conducted seemingly similar experiments seeking to improve treatments for wounded soldiers, or for pilots downed in freezing waters, or with respect to diseases like malaria. Prosecutors at the Nuremberg Trials, therefore, had to condemn the presumptively "immoral" experiments conducted by German researchers without in the same breath indicting seemingly similar experiments conducted by American or British scientists.

Mellanby had parallel concerns. In the 1973 edition of his book he had to differentiate his own experiments on COs, which he claims were "performed in a reasonably ethical manner," (16) in contrast to the treatment of "the victims of the [German] experiments" (Mellanby 1973, 16). He did so by arguing that his COs had "taken part as volunteers" (16)—a loosely defined concept that appears to mean that his subjects had a choice of other ways of performing alternative service. Whereas, by contrast, the Nazi scientists performed their experiments on "prisoners, compelled to take part in work which they knew would quite probably mean their death, and if sabotage was possible they looked upon it as their duty" (Mellanby 1947, 150).

Yet Mellanby also recognized that some of the Nazi experiments on battlefield injuries and on diseases (like malaria and typhoid) could prove valuable and he sought to save the data from these experiments. In pursuit of this objective he drew on his social connections to have himself appointed as the BMJ's first foreign correspondent. He then used his newly minted credentials to gain admission to the Nuremberg Doctors' Trial as a journalist (Weindling 2004, 207) and he subsequently published an article on the trial in the BMJ. He also commented on the trials in the 1973 edition of *Human guinea-pigs*. In these commentaries, and in an earlier 1946 letter published in *The Lancet*, Mellanby defended the German researchers who conducted scientific experiments on curing diseases. He also cast aspersions on their American prosecutors, characterizing them as a "monoglot" group of "attorneys [who] lounged about and sat with their feet on the table, chewing gum and apparently trying to simulate an out-of-date courtroom movie" (Mellanby 1973, 185). Mellanby says that the German defense lawyers, in contrast, make "a somewhat better impression" (1973, 185). Mellanby also parroted the German defense lawyers' *Tu Quoque Defense*, a "what-about-what-you-did" charge leveled at the Americans. As Mellanby parses it, how dare the Americans level these charges against Germans when "[in] America criminals in gaol [jail] did take part in tests involving malaria infection not unlike those done at Dachau by [Dr. Klaus Karl] Schilling [1871–1946]—and…. In view of this fact, it is somewhat ironical that Schilling was…found guilty of mass murder, and summarily hanged" (Mellanby 1973, 190).

Mellanby's claim that Americans were as guilty of unethical experimentation as the Germans was based on experiments with anti-malarial drugs conducted by the University of Chicago's Malaria Research Unit's at Jolliet-Stateville prison from 1944 to 1946, in which researchers infected 500 healthy white male prison inmates with malaria to test the efficacy of various anti-malaria drugs (Alving et al. 2013, 548–567). Although superficially similar to Schilling's experiments, as was pointed out at the trial, the Jolliet-Stateville experiments were conducted quite differently. One difference was that no deaths were associated with the Jolliet-Stateville experiments[9]; whereas, according to such witnesses as Schilling's assistant, Czech prisoner Dr. Franz Blaha, hundreds of deaths were associated with Schilling's Dachau experiments. Here is a portion of Dr. Blaha's (1946) testimony (as officially transcribed into English).

> During my time at Dachau I was familiar with many kinds of medical experiments carried on there with human victims. These persons were never volunteers but were forced to submit to such acts. Malaria experiments on about 1200 people were conducted by Dr. Klaus Schilling between 1941 and 1945… The victims were either bitten by mosquitoes or given injections of malaria sporozoites taken from mosquitoes. Different kinds of treatment were applied, including quinine, pyrifer, neosalvarsan, antipyrin, pyramidon and a drug called 2516 Behring. I performed autopsies on bodies of people who died from these malaria experiments. Thirty or forty died from the malaria itself. Three hundred to four hundred died later from diseases which proved fatal because of the physical condition resulting from the malaria attacks. In addition there were deaths resulting from poisoning due to overdoses of neosalvarsan and pyramidon. Dr. Schilling was present at the time of my autopsies on the bodies of his patients.

Notice that Dr. Blaha testified that none of Schilling subjects were volunteers.[10] Several other witnesses confirmed Dr. Blaha's claims. In contrast, all of the subjects

---

[9] One inmate died during the study; although researchers claim this inmate's death was not a result of their experiment, a memoir by fellow inmate Nathan Leopold indicates that the he believed the death was associated with the experiment (Leopold 1974 [1958], 320). See also Miller (2013).

[10] Furthermore, unlike Schilling's experiments, the University of Chicago's Jolliet-Stateville malaria experiments were publicized in a photojournalist story in *Life* magazine that included photographs of un-protesting prisoners receiving malarial mosquito bites. It was the pictures on *Life Magazine* that alerted the German defense lawyers to the Jolliet-Stateville study. See Anonymous (1945, 44).

in the Jolliet-Stateville experiment signed the following form attesting to their volunteer status.

> I..., N[umber]. ...aged..., hereby declare that I have read and clearly understood the above notice [describing the experiments and associated risks], as testified by my signature below, and I hereby appeal to the University of Chicago, which is at present engaged in malarial research at the orders of the Government, for participation in the investigation of the life-cycle of the malarial parasite. I hereby connect all risks connected with the experiment and on behalf of my heirs and my personal and legal representatives assume all the risks of this experiment. I hereby absolve from such liability the University of Chicago and all the technicians and researchers who take part in the above-mentioned investigations. I similarly absolve the Government of the State of Illinois, The Director of Department of Public Security of the State of Illinois, the warden of the State Penitentiary at Joliet-Stateville, and all employees of the above institutions and Departments, from all responsibility, as well as from all claims and proceedings or Equity pleas, from any injury or malady, fatal or otherwise, which may ensue from these experiments.
>
> I hereby certify that this offer is made voluntarily and without compulsion. I have been instructed that if my offer is accepted I shall be entitled to remuneration amounting to [100] dollars, payable as provided in the above notice (Green 1948, 457).[11]

Based on the contrast in mortality outcomes and the presence of signed consent forms in the one case and their absence in the other, the Nuremberg Tribunal dismissed the German defense's claims of similarity between the Dachau and Jolliet-Stateville experiments. In 1948 Illinois Governor Dwight Green (1897–1958) commissioned a review of the Jolliet-Stateville malaria study (Green 1948, 457).[12] Andrew C. Ivy (1893–1978), a consultant/contributor to the document known as the "Nuremberg Code," headed the commission.[13] After assessing the Jolliet-Stateville experiments in terms of criteria for ethical experiments that Ivy had recommended to the Nuremberg Tribunal,[14] the committee cited as "An example of human

[11] For a slightly different version of the form, see Harcourt (2011, 443).

[12] The central question the commission addressed was whether a parole board should consider a prisoners' participation in an experiment when decided about whether to parole a prisoner.

[13] Ivy's role in formulating the document now known as the Nuremberg Code is detailed in historian Paul Weindling's authoritative study, *Nazi Medicine and the Nuremberg Trials: From Medical War Crimes to Informed Consent* (2004). A helpful table on pages 357 to 358, "The Evolution of the Nuremberg Code," charts the respective contributions to the Code's ten rules by each of the tribunal's two consultants, Andrew Ivy and Leo Alexander.

[14] Ivy's three criteria for ethical experiments are: 1. Consent of the human subject must be obtained. All subjects have been volunteers in the absence of coercion in any form. Before volunteering, the subjects have been informed of the hazards, if any. 2. The experiment to be performed must be based on the results of animal experimentation and on a knowledge of the natural history of the disease under study and must be so designed that the anticipated results will justify the performance of the experiment. The experiment must be such as to yield results unprocurable by other methods of study which are necessary for the good of society. 3. The experiment must be conducted (a) only by scientifically qualified persons and (b) so as to avoid all unnecessary physical and mental suffering and injury and (c) only after the results of adequate animal experimentation have eliminated any a priori reason to believe that death or disabling injury will occur. If there is any a priori reason that accidental death or disabling injury may occur, as in such experiments as those of Walter Reed in which the mosquito was demonstrated to transmit yellow fever, then medi-

experiments which were ideal because of their conformity with the [Ivy] ethical rules the [malaria] experiments at Stateville" (Green 1948, 457).

A 2013 retrospective review by philosophically trained NIH bioethicist Franklin Miller reached a similar conclusion. Pivotal to Miller's (2013) analysis were autobiographical comments by an infamous Jolliet-Stateville prisoner who participated in the malaria experiment: Nietzschean child-murderer Nathan Leopold (1904–1971). Leopold was incensed by the German defense lawyers' claim that the Jolliet-Stateville prisoners' participation was "involuntary." "That was absolutely false," Leopold (1958, 206) wrote in his 1958 autobiography, *Life Plus 99 Years*. "The young docs and Dr. Alving leaned over backward in handling the matter of volunteering in a scrupulous and ethical manner" (306). Leopold (1958) notes that he and all other Inmates were informed of the experiment by "an announcement [that] was read over the institutional radio, asking for volunteers. The first day 487 men volunteered" (307). Leopold also wrote that

> The docs explained in great detail to each and every volunteer before he was used just what it was planned to do. We were told that there was danger, that we might be sick, that we might die. No man was coerced or even persuaded. If anything, the Army officers threw their weight the other way. Every man who went on the project at Stateville did so because he wanted to, almost because he insisted on it. The problem was never that of finding volunteers; rather it was the difficult one of selecting among the large number who wanted to be subjects.... I really had to twist... arms to get my own chance to go on the project (306).

Based on statements like these, Miller concluded that, "the voluntariness of consent...[in] principle was satisfied in the recruitment and consent process for the Stateville malaria experiments. Although we have no record of exactly what the prospective subjects were told about the nature of the experiments and what to expect, if Leopold's description is accurate, then it appears that they were sufficiently informed to give valid consent" (Miller 2013, 558). What Miller (2013) finds most persuasive was that, "According to Leopold, a major motivating factor for at least some of the prisoners was the opportunity to contribute to research that might aid the war effort. The incentives to participate in the Stateville malaria experiments seem far from irresistible, especially as many prisoners did not volunteer" (556).

To return to the issue of Mellanby's attitude toward the German scientists: What intrigues me is that given a choice between accounts of the superficially similar American and German malaria experiments, Mellanby condemns the Americans but exonerates the Germans. For example, in the 1947 BMJ article Mellanby wrote,

---

cal scientists should serve or should have served as volunteers along with nonscientific personnel (Green 1948, 457).

Ivy recommended these three rules to the Nuremberg Tribunal, and in a report that he sent to the AMA, and to the US Army (Ivy, circa 1946). The AMA adopted a simplified version of these guidelines in 1946 as a *pro forma* gesture distinguishing "ethical" American experiments from their "unethical" Nazi counterparts. The AMA's three requirements were: (1) voluntary consent, (2) prior animal experimentation, and (3) proper medical protection and management. They were published under the AMA Judicial Council's "Ethical Guidelines for Clinical Investigation," for 1946. These criteria were never seriously promulgated or enforced.

> At Dachau in 1946 Dr. Klaus Karl Schilling, the eminent malariologist, was hanged for his part in experiments in which some 1,200 prisoners were infected with malaria. He had a greater claim than most of the experimenters to the title of scientist.... He was over 70 when these experiments were carried out. Believing that the experiments would provide an unequalled opportunity for the advancement of malariology he used a mild strain of benign tertian malaria which would be unlikely to prove fatal in healthy individuals, and he himself appears always to have administered drugs carefully and scientifically. He is said usually to have obtained the consent of the prisoners before he infected them (Mellanby 1947, 148).

In the 1973 edition of *Human guinea-pigs* Mellanby reiterates his denial that Schilling, "a scientist of international reputation, for some years a leading member of the League of Nations Malaria Commission... infected over a thousand prisoners with malaria, and, according to the indictment, several hundred died of the disease. I simply do not believe this [because] a mild strain of Benign Tertian malaria was used, and that this itself would seldom cause death even in patients in a poor state of health" (189). He continues, "those who knew and worked with [Schilling] did not think that he would be a party to serious and inhumane crimes and my investigations in 1946 do not support some of the criticisms made by some other Germans in later years.... There seems little doubt that Schilling explained his work to the participants, and that they were quite genuine volunteers—if it is possible for a volunteer to exist within a prison" (189). Mellanby's (1973) sole criticism of Schilling was that his "work was, scientifically, rather disappointing. He confirmed the known properties of several antimalarial drugs, but did not make a scientific breakthrough" (189).

In a December 7, 1946 (pre-trial) letter to *The Lancet*, Mellanby condemns the idea of destroying data from the Nazi experiments as "pernicious sentimentality" (Mellanby 1946, 850),[15] even as he confesses "a great deal of sympathy for some of those who were responsible for carrying out the experiments.... especially serious research workers" (850). He explains further that, "Given the chance of using prisoners for experiments, which one believed to be of great importance and value to mankind, what would one do if...the victims were dangerous criminals who were anyhow condemned to death and likely to die in some particularly abominable manner?" Mellanby (1946) was "not sure what I myself should have done" (850).[16]

Mellanby (1946) concludes his letter by emphasizing the need to preserve the Nazi data (850). He seems unconcerned by the possibility that publishing data from these experiments could valorize the morally deplorable means used to attain them or lessen the taint associated with those who resorted to these means. Yet Nazi scientists, like Schilling, pleaded for publication of their experiments on precisely the

---

[15] Mellanby also attempts Golden Rule legitimation of using data from the experiments, arguing that "If I myself had been a victim, and some of the results of value or of interest had been obtained from my death, I am sure that I should have preferred to know that this knowledge would have been used and that I had not died entirely for nothing."

[16] He does acknowledge that some of the prisoners might be innocent but argues that the results of the research should be published but, "At all costs sensationalism must be avoided, and it might be perhaps to grade [these reports] as 'confidential' and make them available only to bona-fide investigators" (Mellanby 1946, 850).

grounds that, "It would be an enormous help… for my colleagues, and a good part to rehabilitate myself" (Schilling 1945). To repeat for clarity and emphasis: Schilling is pleading for publication because, as the crude English translation reads, it would do "a good part to rehabilitate myself," i.e., to publish data from the experiment would validate Schilling's experimental methods and rehabilitate the experimenter (Schilling 1945).

Mellanby's sympathies with the Nazi researchers were so strong that he wrote a testimonial defending another defendant at the trials, Professor Gerhard August Heinrich Rose (1896–1992). Rose was accused of supplying the prisoners that were used as human guinea pigs in Schilling's malaria experiments and in the typhus experiments of *Sturmbannführer*, (Waffen SS Major), Dr. Erwin Oskar Ding-Schuler (1912–1945), surgeon and head physician at the Buchenwald Concentration Camp. Ding-Schuler was a long-time Nazi party member who joined the party in 1932, when the Nazis assumed control of the government. He volunteered for the SS in 1936 and became head physician at Buchenwald in 1939. In this capacity he conducted experiments on about 1000 inmates, many of who died—including the Evangelical Protestant Pastor, Paul Robert Schneider (1897–1939).

None of this seemed relevant to Mellanby (1973), who praised Ding-Schuler's research as the "one important and unique piece of medical research which was done in a concentration camp…the study on typhus immunization" (188–189). Thus even though Mellanby (1973) acknowledged that Ding-Schuler "really did use his victims as though they were guinea pigs," (188) he justified Ding-Schuler's experiment on the grounds that "It was a well-conducted scientific experiment… [resulting in a] vaccine…used on… on the German side… [thus] Ding's results were probably quite valuable" (188). Mellanby (1973) argues further that,

> Ding was a reputable scientist, he appeared to think that the deaths caused by his experiments—probably not more than 250—were justified, remembering the five million who died of typhus during and after the previous war. He appears to have tried to use "volunteers," and some of those who took part and survived…. Nevertheless, either his conscience preyed upon him, or he was afraid of the consequences, for he committed suicide and so avoided capture and trial. His own published account, and his laboratory notes, clearly established his guilt, even though his efforts almost certainly saved more lives than his experiments killed—and when he did the work he, like most other experts, expected typhus to be as serious as it had been 25 years earlier, so for every victim of his experiments 20,000 others might have been saved (188–189).

This is an amazing argument. Mellanby (1973) is claiming that in principle "a reputable scientist" is "justified" in performing a "well-conducted scientific experiment" on prison inmates in "which not more than 250 people died," if "though his efforts [he] almost certainly saved more lives than his experiments killed." One commentator characterized Mellanby's principle as "utilitarian" (Weindling 1996, 1467–1470). This is a misnomer. From founding figure Jeremy Bentham (1748–1832), to his most famous disciples, Harriet and John Mill (1807–1858, and 1806–1873 respectively), the classical utilitarians would never have countenanced such a proposal. Bentham initially deployed his Greater Happiness Principle to critique the conservative views of William Blackstone (1723–1780) and others who justified acceptance of inherited custom, practices, and laws even were though they were unresponsive to the emerging needs of a rapidly industrializing and urbanizing society (Blackstone 1765, 120–141). The genius of the classical utilitarian movement was that it boiled down an elitist eighteenth-century morality of moral sensibility to two fundamentals—pain and pleasure—and married these to a form of moral book keeping, the "greatest happiness principle," which was designed to promote happiness and eliminate misery for the greatest number of sentient creatures. The initial targets of the classical utilitarian critique were animal abuse (e.g., bull-baiting), environmental degradation, mass incarceration, religious bigotry, sexism, superstition, slavery, unrepresentative governance, and unsanitary living and working conditions. Bentham, in particular, was profoundly skeptical of elites' claims to entitlement and would never have accepted Mellanby's suggestion that "reputable scientists" should have untrammeled license to experiment on prisoners if, to their minds, the anticipated mortality of the experimental subjects did not exceed the mortality of the disease. Bentham would certainly have rejected such a proposal on the grounds that it is irrelevant to the deterrent, rehabilitative, and morally educative utilitarian justifications for punishment. He would also question whether a captive population in a totally controlled environment would be in a position to "volunteer," and would undoubtedly suggest that experiments, being by their very nature, "experimental," would not typically yield significant results. (As Mellanby himself reported, "no great medical advances were likely to appear as a result of the sufferings of the victims of [Nazi] experiments. Little of the work had been properly planned, few of the investigators were competent, there was a lot of very inaccurate recording and even some deliberate falsification of results. Many of the so-called medical experiments were quite unworthy of that name, and were merely excuses for bestial cruelty" (Mellanby 1973, 186–187). Finally, Bentham would have distrusted any moral account resting on the subjective impressions of an elite, i.e., "reputable" scientists. The thrust of Bentham's critique of Blackstone and his followers was that, as an elite, they interpreted morality and law to favor their own privileged status. The fundamental point of the utilitarian moral calculus was to render moral reasoning inter-subjective, transparent, and thus publicly accountable.

The question remains: since Mellanby understood that innocent people endured enormous suffering at the hands of Nazi doctors, and that their suffering yielding little that was scientifically significant, why did he go out of his way to defend Ding-Schuler, Gerhard Rose, Klaus Schilling, and the other Nazi scientists? Several factors seem to be in play. In his pioneering history, *The Making of British Bioethics*, Duncan Wilson argues that British bioethics was invented to rectify the abuses of "club regulation," an informal system of self-regulation by members of the British medical establishment who ran the profession as an old boys' club with unwritten rules that protected and advanced the interests of fellow gentlemen, like themselves, i.e., men who went to the right schools, who talked the right way, and who knew instinctively what was and what was not "the done thing" (Wilson 2014). It is likely that, to Mellanby's (1973) way of thinking, Schilling, "a scientist of international reputation, for some years a leading member of the League of Nations Malaria Commission," (189) was an honorable member of the club. Consequently, he must be innocent because fellow club members "who knew and worked with him did not think that he would be a party to serious and inhumane crimes" (Mellanby 1973, 189). As for Rose, Mellanby (1973) wrote that he was

> a patriotic German with right-wing views who preferred the effects of the Nazi party on the international standing of Germany to the acceptance of disabilities that the Treaty of Versailles had inflicted, yet I was convinced that he was not a monster. His offence was simply that of supplying cultures of insects to a scientific colleague. I knew from my own experience that this could have been done quite innocently. .... But even if Rose had known exactly what Schilling was doing, he would have had no reason to be concerned, as the experiments were planned in a humane and reasonably non-dangerous manner. I tactlessly said this to a German journalist at the trial, and he reported the matter to Rose's counsel. I was asked to give written evidence on the subject, which I did. Professor Rose was eventually set free. I think that it was absurd that he was ever charged (190–191).

What Mellanby neglects to mention was that Rose did more than supply mosquitos for experiments, he also supplied consignments of prisoners to Ding-Schuler and other researchers to serve as human subjects in experiments that were never planned to be "humane" and that were executed in an atrociously inhumane manner. Rose, moreover, was unrepentant about his role in the Holocaust and would later argue that the Nuremberg Trials were "*Unrechts-Justiz*" (unfair justice). He attempted, unsuccessfully, to have the Nuremberg verdicts against himself, and other Nazi scientists, quashed (Weindling 2004, 335).

Mellanby sympathized with Rose and Schilling as "Good Old Boys," eminent members of a gentleman's club who, by virtue of their age and stature, were above reproach. Ding-Schuler, however, was not a member of the club, and so the question arises: why did Mellanby defend him? What Mellanby emphasizes is Ding-Schuler's 1943 publication of his vaccine experiments with concentration camp inmates in a well-respected international journal, *Zeitschrift für Hygiene und Infectionskrankheiten* (*Journal of Hygiene and Infectious Diseases*, 1891–1961). As it happens, a year earlier, in 1942, Mellanby himself sought to develop a typhus vaccine by means of a similar set of life-endangering experiments using his COs as human guinea-pigs. When he sent his proposal to his uncle, Edward Mellanby, Secretary of the MRC, it was rejected, not on ethical grounds, but because it might generate "'ballyhoo'... and unavoidable [bad] publicity;" (Weindling 1996, 1468) and also, more pragmatically, because Americans had already developed and were testing a typhus vaccine (Weindling 1996, 1468). Since Mellanby himself had wanted to perform Ding-Schuler type experiments on his own human guinea-pigs, it is not unreasonable to suggest that he likely rationalized it in precisely the same pseudo-utilitarian terms that he attributed to Ding-Schuler. Consequently, to admit Ding-Schuler's guilt was to accept that, save for the intervention of Uncle Edward, he himself would have been as guilty of crimes against humanity as was Ding-Schuler.

One might also speculate that Mellanby's felt comfortable with Ding-Schuler's experiments, and could propose performing similar typhus experiments on his COs because of his scientific training as an entomologist. Entomologists are scientists who study insects, not humans. Unlike Pappworth's scathingly ironical title "Human Guinea Pigs," Mellanby's "Human guinea-pigs" does not metaphorically herald a critique of researchers' inhuman treatment of humans as if they were mere lab rats. Instead it uncritically describes the curious incident of COs being used as lab rats. As Mellanby remarks at one point that since he had done "some work on vitamin A deficiency in rats, I wondered if similar experiments … could be carried out on the [human] volunteers" (Mellanby 1973, 129–130). In another place, Mellanby notes that he was "interested in questions of 'water balance, [and] Though I had not studied this subject before in man, my work for some years had been concerned with not dissimilar problems as they concerned insects, and it was therefore very interesting to see how the same factors operated in these two completely different forms of life" (Mellanby 1973, 125). Mellanby (1973) writes as if his COs were just a convenient sample of humanoid forms of life as interesting to study as insects or rats—although, as he put this point elsewhere, more challenging to "manage" (52).

Mellanby never mentions the principles that the Nuremberg tribunal used to distinguish legally and morally permissible experiments on humans from the illegal and morally impermissible experiments conducted by the Nazi researchers: which later became known as the Nuremberg Code (*US vs Karl Brandt,* 181–182). The first and most famous of these principles states, "the voluntary consent of the human subject," is a prerequisite of moral and legal experiments on humans (*US vs Karl Brandt,* 181–182). This principle requires that researchers make "known to [a potential experimental subject] the nature, duration, and purpose of the experiment; the method and means by which it is to be conducted; all inconveniences and hazards reasonably to be expected; and the effects upon his health or person, which may possibly come from his participation in the experiment." (*US vs Karl Brandt,* 181–182). Thus, insofar as Mellanby took cognizance of these principles, by the time he issued the second edition in 1973, he must have been aware that he violated the principle of informed consent. Recall that he admits to deliberately withholding disturbing information from his "volunteers" (Mellanby 1973, 54). Had he paid attention, he might also have noticed that the tribunal stipulated that "The experiment should be so conducted as to avoid all unnecessary physical and mental suffering and injury," and require "the scientist in charge… to terminate the experiment at any stage, if he has probable cause to believe… that a continuation of the experiment is likely to result in injury, disability, or death to the experimental subject" (*US vs Karl Brandt,* 181–182). Yet Mellanby and other researchers at his institute conducted dietetic and water deprivation experiments that inflicted unnecessary hardships on their subjects, that led to permanent injuries and temporary night blindness and that, in one case, led to a person's death (Weindling 1996, 1468; Pemberton 2006, 557). Lastly, the fact that the tribunal required that "the experiment should be conducted only by scientifically qualified persons," should have given Mellanby pause, since it is questionable whether an entomologist, an expert on insects, is qualified to design and supervise experiments on humans. One suspects that consciously, or perhaps, subconsciously, Mellanby's defense of the German researchers derives, in part, from some sense that, insofar as Ding-Schuler, Rose and Schilling were guilty of abusing human subjects, he too might be subject to these accusations.

Pappworth, Mellanby and the Equanimity of the Establishment

What Mellanby took away from the Nuremberg Doctor's Trial was a sense of frustration and outrage: Frustration that so much of the data from the German concentration camp experiments was lost or worthless, and outrage at the tribunal's treatment of eminent German scientists, who, in his mind, were being unfairly punished for experiments that their American counterparts had conducted with impunity. The trials also influenced Pappworth. He too took note of the similarities between American, British, and German wartime research on diseases and battlefield injuries—he includes the Jolliet-Stateville prison experiments in his list of unethical experiments (Pappworth 1967a, b, 61–62)—but what engaged Pappworth's attention was a feature of the trials never mentioned by Mellanby: the tribunal's criteria for legally and ethically permissible experiments (developed with the assistance of two American consultants, Andrew Ivy (1893–1978) and Leo Alexander (1905–1985)). He also recognized that by convicting German scientists of illegal experiments, the tribunal had forged a touchstone of immorality against which all other experiments could be measured—including those conducted in Britain and America.

The rest of the world learned no such lessons. With the almost immediate onset of the Cold War between the West and the Soviet Block (1945–1990), the world surrendered to retrograde amnesia about Nazi doctors and the Nuremberg trials. As historian Duncan Wilson observes, what later came to be called "the Nuremberg Code…was routinely ignored by researchers in Britain…and elsewhere, who believed the guidelines were designed to prosecute 'barbarians' and did not apply to them" (Wilson 2014, 44). American historian David Rothman (1991) concurs, reporting that, "Neither the horrors described at the Nuremberg Trial nor the ethical principles that emerged from it had a significant impact on the American research establishment. The trial itself did not receive extensive press coverage" (62). Additionally, Rothman (1991) observes, "well into the 1960s, the American research community considered the Nuremberg findings, and the Nuremberg Code, irrelevant to its own work:" (31) Much the same can be said of the British research community.

Yet in the 1960s, Pappworth and an American pen-pal, Harry Unangst of Peck, Kansas, (Gaw 2012, 150–155) would remind governments, the public, and fellow physicians, about the Nazi doctors. Like Pappworth, Unangst was a veteran who served in field hospitals during World War II. And, also like Pappworth—and unlike Mellanby—Unangst was not born with family connections or a proverbial silver spoon in his mouth: he inherited an undistinguished family name, an unfashionable birth place, a minority religious background (Evangelical Revivalism) and personal predilections that should have relegated him to obscure professional outposts. Yet on the verge of applying to Harvard Medical School, Unangst, a Midwestern hick working his way through college, changed his name to "Henry Knowles Beecher," thereby altering his predestined fate. After graduating from Harvard, Unangst, now known as "Beecher," climbed the professional ranks to hold the Henry Isaiah Dorr professorship at Harvard Medical School.

As medical historian Paul J. Edelson (2004) observed, when Pappworth and Unangst-Beecher discovered each other they formed a mutual support team (219–234). Yet their styles differed. Smarting from a lifetime of anti-Semitic rejections, Pappworth believed sinners should be held accountable, so he named names, shamed institutions, and seem to relish speaking truths to power. When the powerful tried shushing him, he replied, "those who dirty the linen and not those who wash it should be criticized. Some do not wash dirty linen in public or private and the dirt is left to accumulate until it stinks" (Pappworth 1990, 1456). Beecher focused instead on reforming the system rather than shaming those who had once been miscreants. This may be because he was aware that he himself was not above reproach. In an unpublished 1965 draft of a paper he admitted administering "unusual tests [with] no adequate explanation … to the subjects, nor was their consent adequately obtained" (Lederer 2016, 28). In the public version of his paper he confessed that he was "'obliged to say that, in years gone by, work in my laboratory could have been criticized' on ethical grounds" (Harkness et al. 2001, 365). Beecher may also have felt guilty about drug experiments he conducted for US Central Intelligence Agency (CIA, founded 1947) in the early years of The Cold War (McCoy 2007, 401–417; Moreno 2016, 107–121). The quest for redemption is a powerful motive. Given his revivalist evangelical background, Beecher must have felt something akin to another repentant revivalist, John Newton (1725–1808), a slave trader turned abolitionist. Newton voiced his repentance in his famous sermon-poem-song, *Amazing Grace*, crying that he "once was lost…was blind, but now he sees" (Rogers 1996) and having seen, became a reformer (Newton 1788). Beecher, as a repentant sinner who daily read a passage from the Bible (Harkness 1999, 465–467), sought redemption as a Christian and a researcher by publishing data on unethical medical experiments and by seeking to right a system that allowed him and others to go astray.

Sharing overlapping objectives, the two reformers began a trans-Atlantic correspondence exchanging lists and details of unethical experiments. Initially the format of their critiques was similar: lists of experiments deemed unethical because they were unconsented, or harmful, or deadly. Both attempted to construct histories of research ethics in which they redubbed Nuremberg Tribunal's principles "The Nuremberg Code" (Beecher 1970, 213–234; Pappworth 1967a, b, 185–212).[17,18] However, their primary audiences differed: Beecher's audience was fellow professionals; Pappworth, having been denied professional platforms, addressed a lay audience. Both whistleblowers faced challenges as they sought to publish, in part because their long lists of "unethical" experiments did not fit nicely into word limits,

[17] Note that Beecher's critique of the Nuremberg Code reflects his background as a researcher and is designed to alter the code to make it useable to fellow researchers. Note further that in this history a set of observations in a legal decision is converted into a moral code denominated the "Nuremberg Code." See also note 18.

[18] Note that in this history, as in Beecher's (see note 17) a set of observations in a legal decision is converted into a moral code denominated the "Nuremberg Code."

but mainly because a lifetime's experience of reading, writing, and teaching medicine provided neither the narrative structures, nor the terminology, nor the background knowledge conducive to moral critiques. After multiple rejections Beecher's (1966) article was eventually published in the *New England Journal of Medicine*—and then only because the journal's editor, Joseph Garland (1893–1973), pruned the article down from 50 to 22 cases (Lederer 2016, 31), changed the paper's title, overrode external reviewers' negative comments, and defied his own board. (Garland retired the year after publishing Beecher's article.)

Pappworth's manuscript suffered a similar fate. It was rejected half a dozen times, in part because as one publisher "put it bluntly, it just isn't, in its present form, a book" (Seldon 2017, 124). As a different manuscript reviewer remarked, "the...hundreds of baldly reported technically worded case histories [would be] virtually unreadable...to the great majority of laymen," i.e., to Pappworth's primary audience (Seldon 2017, 125). Moreover, since, unlike Beecher, Pappworth insisted on revealing the names and institutional affiliations of the researchers whose experiments he deemed "unethical," he made himself and his publisher vulnerable to liable suits. In the end, however, Routledge published his book.

Whatever Pappworth book's flaws as literature, it drew the attention of internationally renowned reviewers, including the moral philosopher Sir Bernard Williams (1929–2003) and the author Arthur Koestler, CBE (1905–1983). It instantly commanded headlines in newspapers from broadsheets like the *Guardian* and the *Times*, to tabloids like the *Daily Express* and *The Daily Mail.* The headlines of the *Daily Express* screamed, "What are your chances of being a human guinea pig?"; the *Daily Mail* stated more sedately, "children, pregnant women, mental defectives, and old people were being used in National Health hospitals as 'guinea pigs' without consent... [Pappworth] likens some of the experiments to research carried out in the Nazi concentration camps" (Seldon 2017, 145–146). Debates erupted in the medical press and in parliament and the good old boys in the medical establishment reluctantly responded by taking some steps to initiate peer review; however, major reforms in Anglo-American research ethics really emerged two decades later with the forging of the lawyer-philosopher-physician/nurse-theologian alliance that came to be known as "bioethics."

During the period when Pappworth's book was commanding center stage Mellanby was no longer experimenting on human subjects. At this point in his life he had established himself as director of a small ecological institute, the Monks Wood Research Station, and had turned his attention to pesticides and the environment. He seemed oblivious to the brouhaha swirling around Pappworth and his American pen pal, Beecher. Then, out of the blue, in 1973, he decided to publish a second edition of *Human guinea-pigs*. It was an act of protest, not against the abuse of research subjects but against a proposed new and more formal system for funding British research that, Mellanby feared, would jeopardize his style of informal improvisational research (Mellanby 1973, 13). The proposed new system, Mellanby argued, would have thwarted his innovations in the prevention and treatment of scabies since, he feared, it would no longer be possible for "a very junior researcher [as Mellanby was at that time, who] thought that something should be done, and then put forward a harebrained scheme which, miraculously [or, since Mellanby's uncle was on the research committee, perhaps not so miraculously] attracted a modest amount of support, even if this only became operational after he crusaded for this support with missionary zeal. ...I must take some credit, as I took responsibility for trying to get my results applied as widely as possible" (Mellanby 1973, 13).

Mellanby also claimed that his book was newly relevant because the issue of conscientious objectors had been revived in the context of the American war in Vietnam. British peace activist Margaret Melicharova challenged Mellanby's claim to be concerned about COs. She concluded that Mellanby's real purpose in republishing *Human guinea-pigs* in 1973, a quarter-of-a-century after it went out of print, was to give him a platform to protect good old boys' club self-regulation from the prospect of external regulation. "Dr. Kenneth Mellanby," she wrote, "was one of many researchers wanting freedom from interference in their projects.... Such scientists are also natural autocrats. Mellanby was convinced that in experimental work... 'it is best for all concerned for one person to be in the position to make every decision that is required'. That 'one person' was, of course, himself" (Melicharova 2006).

As Melicharova observes further, Mellanby's autocratic reign over his human guinea-pigs was not entirely benign. He failed to mention or memorialize the CO who died in his research center, and, she could have remarked, he also neglected to acknowledge COs who suffered from night blindness, who were scarred permanently, and who "contracted major illnesses—TB, impetigo, and migraine, one developed an irregular heartbeat, and another suffered the collapse of capillaries in his knees" (Anonymous 2006, 558–560). As Melicharova might put the point today, Mellanby mentions COs as clickbait to gain an audience for his protests against "the processes of democratising and modernising medicine [that] posed a threat to the individual freedoms of clinical researchers;" (Melicharova 2006) i.e., the cozy system of club regulation that tolerated the treatment of experimental subjects as "human guinea-pigs" who could be blinded, maimed, scarred, and sickened at will, as long as it did not generate a "'ballyhoo'...and unavoidable [bad] publicity" (Weindling 1996, 1468).

So why should anyone in the twenty first century read Mellanby's book? His world was not ours. Today research on animal and human subjects is regulated by ethics committees; today we celebrate the Nuremberg Code; today we hail Beecher and Pappworth as heroic whistleblowers; today international and national organizations issue codes and guidelines on research ethics; and, today, journals and books on bioethics fill libraries and bioethical organizations span the globe. So why read a book so dated that it would not even qualify as yesterday's news? Because, also today, "an estimated 350 studies in the social sciences...have critiqued [research] ethics review systems, especially in Anglophone countries, [not to mention] a number of PhD theses," blogs and LISTSERVs (van den Hoonard 2016, 80). These critics, mindful of the fact that research ethics scandals no longer headline the LA, London, or New York *Times*, or the *Guardian,* or the *Toronto Star* or *The Globe and Mail*, observe that researchers in the English-speaking cultural sphere are not Nazis conducting eugenic, racial, or thanatological research. Moreover, we are well past the era of Tuskegee, Guatemala, and Porton Downs. Consequently, they feel free to evince a resurgent nostalgia for the good old days when researchers were unfettered by bureaucratic oversight, unburdened by niggling technicalities, or by seemingly mindless delays waiting for IRB/research ethics committee's approval. Their ever-louder cries urge us to loosen regulation: to free researchers to focus on research.

Let me close this essay with a comment on a quotation that prefaces it. In letters he wrote when imprisoned as an enemy of Italian fascism, Antonio Gramsci (1891–1937), whose voice was silenced after his jailers let him die from lack of medical care, observed that people allow bad things to recur because they refuse to acknowledge the past as an indicator of the future. In his own words, "History teaches, but it has no pupils" (Gramsci 1919). We need not speculate about what would happen were we to jettison our present admittedly (and unduly) irksome regulatory régime; we need only peruse the pages of *Human guinea-pigs* and other memoires from the pre-regulatory era to rediscover that, without nettlesome oversight, an unqualified bug specialist could perform research on human subjects without telling them about potential adverse effects or consequences, and without reporting any adverse consequences. Mellanby was not a Nazi: like us, he was appalled by and condemned the Nazi's thanatological and racially motivated experiments on prisoners. Yet it is evident from his memoire that, in his quest for a publication in a major scientific journal (or, perhaps, for the approval of a famous uncle), Mellanby was willing to dissemble and to risk the lives and health of his experimental subjects without seeking their informed consent. The lesson that history teaches is that the more regulations and oversight are relaxed, the more likely it will become that researchers will find a way to inform less, to "consent" the less informed more, and to exploit the vulnerable—as long as they can do so without unwelcome "ballyhoo'…and unavoidable [bad] publicity." History teaches, and with the republication of this short memoire, readers have an opportunity to become its pupils.

Bibliography

Alving, Alf S., Branch Craige Jr., Theodore N. Pullman, et al. 1948. Procedures used at Stateville penitentiary for the testing of potential antimalarial agents. *Journal of Clinical Investigation* 27 (3): 2–5. https://doi.org/10.1172/JCI101956.

AMA Judicial Council. 1946. *Ethical guidelines for clinical investigation*. Chicago: American Medical Association.

Anonymous. 1831a, November 5. Editor's reply to Dr. Ryan. *The Lancet* 1(1):224–227.

———. 1831b, November 5. Review of *A Manual of Medical Jurisprudence, compiled from the best Medical and Legal Works &c, &c, being an Analysis of a Course of Lectures on Forensic Medicine &c* by Michael Ryan, M.D. &c. &c. *The Lancet* 1(1):137–143.

———. 1945. Prison malaria; convicts expose themselves to disease so doctors can study it. *Life Magazine*, June. https://books.google.com/books?id=h0gEAAAAMBAJ&pg=PA43&lpg=PA43&dq=Prison+malaria;+convicts+expose+themselves+to+disease+so+doctors+can+study+it.&source=bl&ots=RcmbOYMzd1&sig=2WrKyi-xWkS1Astv3IdXHCRoc8&hl=en&sa=X&ved=2ahUKEwj5-eTm5YvdAhUkiOAKHVBjBKUQ6AEwAXoECAkQAQ#v=onepage&q=Prison%20

malaria%3B%20convicts%20expose%20themselves%20to%20disease%20so%20doctors%20can%20study%20it.&f=false. Accessed 26 Aug 2018.
———. 2006, April 17. Commentary: Guinea-pigs' private war. *Journal of Epidemiology* 35:558–560. https://doi.org/10.1093/ije/dyl031. First published in the *Sheffield Telegraph*, August 29, 2003, 12.
Beecher, Henry K. 1966. Ethics and clinical research. *New England Journal of Medicine* 274: 1354–1360. https://doi.org/10.1056/NEJM196606162742405.
Beecher, Henry. 1970. *Research and the individual: Human studies*. Boston: Little, Brown and Company.
Benigno, Antonio. 1900. *Informed consent agreement*, trans. English from Spanish (Rector and Visitors of the University of Virginia, November 26. © 1998–2004).
Blackstone, William. 1765. *Commentaries on the Laws of England*. Oxford: Clarendon Press, Book I, Chapter I. Commentaries 1:120–141. http://press-pubs.uchicago.edu/founders/documents/amendIXs1.html. Accessed 14 May 2018.
Blaha, Franz. 1946. Testimony of Dr. Franz Blaha Document Number 3249-PS, Exhibit USA-66 January 11, 1946," *Nuremberg Trial Proceedings* Vol. 5 *THIRTY-SECOND DAY, Friday, 11 January 1946 Morning Session* http://avalon.law.yale.edu/imt/01-11-46.asp. Accessed 25 Aug 2018.
Booth, Christopher. 1994, December 10. Obituary: M. H. Pappworth. *British Medical Journal* 309:1577–1578. https://doi.org/10.1136/bmj.309.6968.1577.
Brody, Howard, Zahara Meghani, and Kimberly Greenwald, eds. 2009. *Michael Ryan's writings on medical ethics*. Dordrecht: Springer.
Carroll, James (1906). *Report to the surgeon general*. Philip D. Hench Walter Reed Collection, University of Virginia, August 18. http://yellowfever.lib.virginia.edu/reed/collection.html.> Accessed 4 Aug 2017.
Council of International Organization of Medical Sciences (CIOMS) in Collaboration with the World Health Organization. 2016. *International ethical guidelines for health-related research involving humans*. Geneva: CIOMS.
Duke of York's Laws. 1665–1675. https://www.nycourts.gov/history/legal-history-new-york/documents/charters-duke-transcript.pdf. Accessed 9 Aug 2018.
Edelson, Paul J. 1994. Adopting Osler's 'principles': Medical textbooks in American medical schools, 1891–1906. *Bulletin of the History of Medicine* 68 (1): 67–84.
———. 2004. Henry K. Beecher and Maurice Pappworth: Honor in the ethics of human subjects experimentation. In *Twentieth century ethics of human subjects research: Historical perspectives on values, practices, and regulations*, ed. Volker Roelcke and Giovanne Maio, 219–234. Stuttgart: Franz Steiner Verlag.
Franklyn, Perring. 1994. Obituary: Kenneth Mellan by. *The Independent*, Tuesday, July 11. https://www.independent.co.uk/news/people/obituary-kenneth-mellanby-1406138.html. Accessed 6 Aug 2018.
Gaw, Allan. 2012. Exposing unethical human research: The transatlantic correspondence of Beecher and Pappworth. *Annals of Internal Medicine* 156: 150–155.
Gisborne, Thomas. 1794. *An enquiry into the duties of men in the higher and middle classes of society in Great Britain resulting from their respective stations, professions and employment*. London: B. and J. White.
Glynn, Ian, and Jenifer Glynn. 2004. *The life and death of smallpox*. New York: Cambridge University Press.
Gramsci, Antonio. 1919. *Letter from prison, June 21, 1919*. translated by Hamish Henderson. http://en.wikiquote/wiki.Antonio_Gramasci. Accessed 29 July 2018.
Green, Govenor Dwight H. 1948. Ethics governing the service of prisoners as subjects in medical experiments. *Journal of American Medical Association* 136 (7): 457–458. https://doi.org/10.1001/jama.1948.72890240001006.
Harcourt, Bernard E. 2011. Making willing bodies: The University of Chicago human experiments at Stateville Penitentiary. *Journal of Social Research* 78: 443–478.

Harkness, Jon. 1999. Henry Beecher. In *American National Biography Volume 1*, ed. John A. Garraty and Mark C. Carnes. New York: Oxford University Press.
Harkness, Jon, Susan E. Lederer, and Daniel Wikler. 2001. Laying ethical foundations for clinical research. *Bulletin of the World Health Organization* 79 (4): 365.
Hartley, Leslie P. 1953. *The go-between*. London: Hamish Hamilton. https://en.wikiquote.org/wiki/L._P._Hartley. Accessed 8 Aug 2018.
Hazelgrove, Jenny. 2004. British research ethics after the Second World War: The controversy at the British postgraduate medical school, Hammersmith hospital. In *Twentieth century ethics of human subjects research: Historical perspectives on values, practices, and regulations*, ed. Volker Roelcke and Giovanni Maio, 180–197. Stuttgart: Franz Steiner Verlag.
Ivy, Andrew C. (circa 1946) *Report of war crimes of a medical nature committed in Germany and elsewhere on German nationals and the nationals of occupied countries by the Nazi Regieme during world war.* Archives of the American Medical Association, Chicago.
Kaufman, Sharon R. 1997. The World War II plutonium experiments: Contested stories and their lessons for medical research and informed consent. *Culture, Medicine, and Psychiatry* 21 (2): 161–197. https://doi.org/10.1023/A:1005360928209.
Lederer, Susan E. 2008. Walter Reed and the yellow fever experiments. In *The Oxford textbook of clinical research ethics*, ed. Ezekiel Emanuel, Christine Grady, Robert A. Crouch, Reidar Lie, Franklin G. Miller, and David Wendler. New York: Oxford University Press.
———. (2016, Winter). Ethics and clinical research in biographical perspective. *Perspectives in Biology and Medicine* 59(1):18–36.
Leopold, Nathan. 1974[1958]. *Life Plus 99 Years*. Garden City: Greenwood Press.
Maitland, Charle. 1722. *Mr. Maitland's account of inoculating the small pox.* London: Printed for the author, by J. Downing, and to be sold by J. Roberts, at the Oxford-Arms in Warwick-Lane, 1722). http://pds.lib.harvard.edu/pds/view/8072674. Download courtesy of the Harvard University Library, Accessed 3 Oct 2018.
McCoy, Alfred E. 2007. Science in Dachau's shadow: Hebb, Beecher, and the development of CIA psychological torture and modern medical ethics. *Journal of the History of the Behavioral Sciences* 43 (4): 401–417.
McCullough, Laurence. 1998. *John Gregory and the invention of professional medical ethics and the profession of medicine*. Dordrecht/Boston: Kluwer Academic Publishers.
Melicharova, Margaret. 2006, Autumn. The interest of science: Scabies and conscience. *Peace Matters* 52. http://www.ppu.org.uk/coproject/resources/guineapigs.html. Accessed 24 Aug 2018.
Mellanby, Kenneth. 1946, December 7. A moral problem. *The Lancet*, no. ii, 850.
——— (1947, January 25). Medical experiments on human beings in concentration camps in Nazi Germany. *British Medical Journal* 1(4490):148–150.
———. 1973[1945]. *Human guinea-pigs*. London: Merlin Press.
Miller, Franklin G. 2013. The Stateville penitentiary malaria experiments: A case study in retrospective ethical assessment. *Perspectives in Biology and Medicine* 56 (4): 548–567. https://doi.org/10.1353/pbm.2013.0035.
Montagu, Lady Mary. 1763[1718]. *Letters of the right honourable lady M--y W--y M--e: Written during her travels in Europe, Asia and Africa*, Vol. II, Letter XXXIII, To Mrs. S. C., 61–63.
Moreno, Jonathan. 2016, Winter. Acid brothers: Henry Beecher, Timothy Leary and the psychedelic of the century. *Perspectives in Biology and Medicine* 59(1):107–121.
National Commission for the Protection of Human Subject of Biomedical and Behavioral Research. 1978. *The Belmont report: Ethical guidelines for the protection of human subjects of research*. Washington DC: US Government Printing Office, 1978. https://www.hhs.gov/ohrp/regulations-and-policy/belmont-report/index.html. Accessed 21 Sept 2018.
Newton, John. 1788. *Thoughts upon the African slave trade*. London: J. Buckland and J. Johnson. https://archive.org/stream/thoughtsuponafri00newt#page/n3/mode/2up. Accessed 16 July 2018.

Osler, Sir William. 1901. *The principles and practice of medicine*. 4th ed. New York: D. Appleton and Company.
———. 1907. The evolution of the idea of experiment in medicine. *Transactions of the Congress of American Physicians and Surgeons* 7: 7–8.
Pappworth, Maurice Henry. 1962. Human Guinea pigs: A warning. *Twentieth Century Magazine* 171: 66–75.
———. 1967a. *Human Guinea pigs: Experimentation on man*. Boston: Beacon Press. (All references to this edition.).
———. 1967b. *Human Guinea pigs: Experimentation on man*. London: Routledge & Kegan Paul Limited.
Pappworth, M.H. 1990. 'Human Guinea pigs'—A history. *British Medical Journal* 301: 1456–1460.
Pemberton, John. 2006. Medical experiments carried out in Sheffield on conscientious objectors to military service during the 1939–45 war. *International Journal of Epidemiology* 35: 558–560.
Percival, Thomas. 1985[1803]. *Medical Ethics; Or, A Code of Institutes and Precepts, Adopted to the Professional Conduct of Physicians and Surgeons*, edited by Edmund Pellegrino (facsimile edition). Birmingham, AL: The Classics of Medicine Library.
Reed, Walter. 1900, November 22. *Letter from Walter Reed to Emilie Lawrence Reed*, Philip D. Hench Walter Reed Collection, University of Virginia. http://yellowfever.lib.virginia.edu/reed/collection.html. Accessed 22 Sept 2018.
Rogers, Al. 1996. *Amazing Grace: The Story of John Newton*. https://www.anointedlinks.com/amazing_grace.html. Accessed 30 Apr 2018.
Rothman, David. 1991. *Strangers at the bedside: A history of how law and bioethics transformed medical decision making*. New York: Basic Books.
Ryan, Michael. 1831a. *A manual of medical jurisprudence, compiled from the best medical and legal works &c, &c, being an analysis of a course of lectures on forensic medicine &c*. London: Renshaw and Rush.
———. 1831b, November 5. Letter From Dr. Ryan. *The Lancet* 1:224.
Schilling, Klaus Karl. 1945. Dr. Klaus Karl Schilling testifies at the trial of former camp personnel and prisoners from Dachau. United States Holocaust Museum. https://collections.ushmm.org/search/catalog/pa1069345. Accessed 25 Aug 2018.
Seldon, Joanna. 2017. *The whistle-blower: One Man's Battle against the medical establishment*. Buckingham: The University of Buckingham Press.
Sternberg, George. 1900. *Letter to Agramont*. Philip D. Hench Walter Reed Collection, University of Virginia. https://search.lib.virginia.edu/catalog/uva-lib:2513789. Accessed 3 Aug 2017.
US vs Karl Brandt, et al. 1949. In: *Trials of War Criminals before the Nuremberg Military Tribunals under Control Council Law* No. 10, Washington, D.C.: U.S. Government Printing Office, Vol. 2, 181–182. http://www.nus.edu.sg/irb/Articles/Nuremberg.pdf. Accessed 23 Aug 2018).
van den Hoonard, Will C. 2016. Book review; *The Ethics Police: The Struggle to Make Human Research Safe*. *Journal of Empirical Research on Human Research Ethics* 11 (1): 80–83.
Weindling, Paul. 1996. Human Guinea pigs and the ethics of experimentation: The *BMJ*'s correspondent at the Nuremberg medical trial. *British Medical Journal* 313: 1467–1170.
———. 2004. *Nazi medicine and the Nuremberg trials: From medical war crimes to informed consent*. New York: Palgrave Macmillan.
Wilson, Duncan. 2014. *The making of British bioethics*. Manchester: Manchester University Press.
World Medical Association. 1964. *Declaration of Helsinki*. https://www.wma.net/wp-content/uploads/2018/07/DoH-Jun1964.pdf Accessed 8 Aug 2018.

Chapter 19
Human Guinea Pigs: Uncovering Principles for Ethical Research through a Personal Narrative

Ana S. Iltis

> "Can you believe how things used to be?"
> "We've come a long way."
> "We would never do that today."
> "We could never do that today."

Reading historical accounts of patient care and human research practices through the lens of contemporary bioethics can instill a sometimes-unjustified sense of moral superiority and confidence in modern practices. We assure ourselves of the vast differences between then and now, between us and them. Sometimes such accounts also should instill fear. If "people like us" thought they were making good choices and other "people like us" stood by and either supported or ignored various activities, we should assume "people like us" remain capable of poor judgment and grave wrongdoing. In other cases, such accounts can encourage us to re-assess our practices to see where they need improvement even absent egregious ethical violations. It is instructive to understand the assumptions that guided old practices and to examine the common threads that join the past to the present. What drove people? What informed their choices? Do we hold similar assumptions today? What did they not see in themselves, and what are we missing about contemporary practices? What obligations did they attribute to themselves and others, and how did they fulfill them? This is why, for example, the similarities and differences between eugenics practices in the past and present efforts to select for or against characteristics through preimplantation genetic diagnosis or prenatal diagnosis have garnered attention (Mehlman 2011; Iltis 2016; McCabe and McCabe 2011). With the possibility of editing the human genome on the horizon, these inquiries will be ever more

A. S. Iltis (✉)
Department of Philosophy, Wake Forest University, Winston-Salem, NC, USA
e-mail: iltisas@wfu.edu

L. M. Rasmussen (ed.), *Human Guinea Pigs, by Kenneth Mellanby: A Reprint with Commentaries*, Philosophy and Medicine 134,
https://doi.org/10.1007/978-3-030-37697-0_21

important. The reprinting of Kenneth Mellanby's *Human Guinea Pigs* is an important contribution to understanding our past and interrogating our present (see Reverby 2012 on the use of historical events in research ethics to inform contemporary decisions).

Human Guinea Pigs invites us to consider human research practices in the United Kingdom in the early part of the twentieth century, and some of what we read allows us to separate ourselves from history. We read of yet another case of using women (prostitutes, perhaps) to spread disease, for instance, and we imagine that researchers today would not rely on the same approach (p. 46; see also Reverby 2011). We also find a scientist interested in preserving the results of experiments conducted by Nazis on concentration camp victims and who even defended some of the experiments (see Baker, Chapter "*Human guinea-pigs*: Mellanby, Pappworth and Club Regulation" in this volume and LaFleur et al. 2007 for further discussion of these issues.) Yet, *Human Guinea Pigs* also reveals a scientist – Kenneth Mellanby – attuned to many of the ethical concerns typically associated with human research today. Mellanby shows an eagerness to conduct socially valuable, scientifically valid research while respecting and protecting the subjects/participants in his investigation, ensuring that the risks his research posed are appropriate relative to the benefits, and avoiding the unfair use of conscientious objectors. All of these are often-cited obligations for the ethical conduct of research summarized in the literature and captured in numerous guidelines, regulations and codes (see, for example, Emanuel et al. 2000; World Medical Association 2013; CIOMS 2002; ICH 1996). Mellanby grappled with or at least mentioned all these considerations, sometimes implicitly and sometimes explicitly. Whether we agree with his interpretation of these obligations to society and to potential and actual participants is a different matter. His discussion demonstrates both awareness of and concern for these domains despite operating without the benefit of explicit codes, guidelines, requirements, or robust oversight. Mellanby's assessment of the role of social institutions in research and the importance of liberating science from their grips to facilitate progress will speak to researchers today who feel constrained by funding systems and regulatory oversight.

Mellanby's text is particularly compelling in calling us to introspection because *Human Guinea Pigs* reads like a story. He offers a first-person narrative account of decades of planning and carrying out human research. Had we access only to his scientific reports, we likely would have missed out on at least some of these insights. It would have been all too easy to ignore what we share with Mellanby. Most likely, almost no one reading this volume would have paid any attention to his work or even known about it. It would have been relegated to off-site library storage at best. In addition to serving as an important window into the history of human research and inviting us to consider contemporary research practices, the volume serves as a reminder of the importance of narrative in bioethics.

Science and Obligations to Society

Poorly designed or executed biomedical research is unethical because the potential benefits of such research, which include the benefit of knowledge gained, ordinarily will not justify the risks. (The exception here are studies in which the prospect of direct benefit to participants is great enough to justify the research risks alone.) Similarly, if a research question is insignificant, the potential benefit of gaining new knowledge might not justify the risks. This is what gives us the often-cited requirements that research must be valid and have social or scientific value or significance to be ethical (Emanuel et al. 2000; Nuremberg Code 1949). In Mellanby's second chapter, 'Scabies as a Research Problem,' he alludes to these requirements. Understanding the transmission of scabies as well as determining the best treatment for it were of great importance to civilians and the military in his day. Transmission was not well understood, making it difficult to prevent or contain scabies. Treatment posed significant burdens and costs on patients, families, and society. When he decided that the participants might be able to help with dietary research in addition to scabies research, Mellanby started inquiring to determine what kind of dietary research might be most significant or valuable. He considered not only areas of where there were important questions, but which questions would benefit most from the unique circumstances of his research institute. The controlled environment he had created for the scabies experiments and the demonstrated reliability of the participants would facilitate rigorous dietary research, which is notoriously difficult to conduct. The knowledge he set out to gain and the additional lines of inquiry that emerged during his research mattered greatly to society (see his Chap. 6).

The value of research rests in part on the importance of the problem being studied, the study design, fidelity to the protocol, the integrity of data collection, and the quality of reporting. Mellanby's discussion, including his desire to conduct the scabies research in a dedicated controlled facility, reflects a concern with the rigorous design and execution of the study. He notes several times the reliability of the participants and the importance of their adherence to the research requirements for his success. Mellanby's commitments to valid and valuable research prompts the question: can and should we do more to foster valuable, high quality research today? A number of scholars have argued that the answer is a resounding yes. The results of much of the research conducted today go unpublished or studies never are completed (Glasziou et al. 2014; Ross et al. 2013; Ioannidis 2014). Some have argued that many if not most results are false or unreliable (Ioannidis 2005, 2014, 2016; Halpern et al. 2002). Mellanby's focus on ensuring that research is valuable and the results would be useful is a call to identify and address current failures in these areas.

The extent to which research findings will be generalizable to the relevant population also affects the value of research, and this depends partly on the representativeness of the participants. Because Mellanby worried that scabies might "present some special problems" in women, he asked the War Office to set up a facility where women could be treated and scabies could be studied (p. 47).

Contemporary researchers share (or should share) Mellanby's concerns with generalizability and representation. The struggle to make research generalizable to racial and ethnic minorities, women, and children and other populations continues. Today, the recommended solution typically is inclusion in research and sub-group analysis rather than Mellanby's proposal, which was to study women separately. Despite efforts to promote inclusion and foster generalizability, significant problems persist and warrant our attention (Berger et al. 2009; Corbie-Smith et al. 2003, 2004; Mazure and Jones 2015; Hoel et al. 2009; Bennett 1993; Dresser 1992; Shahar 2003).

Obligations to Potential and Enrolled Research Participants

Much of *Human Guinea Pigs* focuses on describing the people who participated in the research, including how they came to be associated with Mellanby, the procedures they underwent, the risks and discomforts they faced, how they were treated in general, and the significance of their dedication and contributions to the successful project. In these participant-focused elements of Mellanby's account, we find evidence of concerns that today we might classify under obligations regarding fair subject selection, favorable risk-benefit ratio, informed consent, and respect for potential and enrolled subjects such as the right to withdraw from research (Emanuel et al. 2000). (We do not see mention of privacy and confidentiality, which likely were in short supply given the living arrangements.)

Mellanby's use of conscientious objectors prompts questions about whether they were required to participate in research in order to be excused from military service, whether they were in some other way pressured to participate, and whether it was unfair to target them. A member of Parliament, Mr. Wilson, asked Mellanby about such issues during his investigation (see p. 72). A visit to the research institute allayed Wilson's fears, according to Mellanby. Mellanby describes the use of conscientious objectors this way: there were "many individuals who felt that they could not take part in the war as combatants and who at the same time wished to serve humanity" (p. 22). He predicted that if they could "alleviate suffering" in a way that "was not solely directed to improving military efficiency," then there likely would be a strong response to a call for research volunteers (p. 23). Mellanby contrasts this with the model of "volunteering" in the military which allows superiors to inform soldiers that they have volunteered. In other cases, soldiers volunteer in hopes of being rewarded (p. 38). Mellanby's participants, he says, "had come entirely of their own free will" and had more lucrative alternatives available to them as conscientious objectors (p. 39). Moreover, they were free to end their participation at any time (p. 39). The extent to which the participants themselves felt free to leave remains unknown.

Today, IRBs, investigators, and scholars understand and apply concepts such as voluntariness, undue influence, and coercion in various ways, yet some common themes emerge. Certain types of research, such as research involving prisoners or phase 1 studies using healthy volunteers, raise special concerns in this regard (see, for example, Fisher et al. 2018; Fisher 2013; Moser et al. 2004; Christopher et al. 2011). In the context of research involving prisoners, the possibility of anything like voluntary informed consent is contested (Edens et al. 2011; Stiles et al. 2012). Debates regarding payments to research participants also often center on voluntariness. Substantive disagreement exists over whether payments ever are permissible and, if so, whether investigators should be limited to covering expenses or may pay participants more (Ackerman 1989; Dickert et al. 2002; Grady 2001). If payments may exceed reimbursement, the purpose and magnitude of those payments is contested (Dickert and Grady 1999). Those who argue for no or limited payments often worry that payments will unduly influence people to participate in research or coerce participation (Macklin 1981; McNeil 1997). Others reject the plausibility of coercive offers, holding that coercion requires a threat (Brody 2002). Yet others argue that we need not worry about undue influence in the context of IRB-approved research (Emanuel 2004). The terms ‘coercion,’ ‘undue influence,’ and ‘voluntariness’ are defined in various ways, leading to different views about the conditions under which the obligation of respect for persons is fulfilled (Nelson and Merz 2002; Appelbaum et al. 2009). Regardless of these differences, the obligation to avoid coercion and undue influence and the requirement that research participation ordinarily be voluntary are widely accepted. Assessing contemporary practices in light of these obligations remains important.

At a number of junctures, Mellanby addresses two concerns about attitudes toward risk that he worried could undermine the long-term interests of society. First, an intolerance of risks could limit research despite the availability of genuine volunteers who “would welcome the opportunity to take similar risks [such as those moutaineers and yachtsmen take] for the benefit of all mankind” (p. 12). This would leave avoidable knowledge gaps. Second, an intolerance of risk, he feared, was undermining the interests of society because, “in the hope of making sure that every new drug is completely safe, almost impossibly stringent standards are being insisted upon” (p. 12). Debates over the permissible level of risk in research (Miller and Joffe 2009), how to assess research risks (Wendler and Miller 2007; Kopelman 2000), how to compare research risks and potential benefits (Weijer and Miller 2004; King and Churchill 2008), and when products should be considered safe-enough for widespread use continue today (Avorn 2005). Debates over these issues in the 1980s in the context of HIV/AIDS research prompted changes at the FDA. HIV activists sought more rapid access to experimental interventions given the death sentence HIV/AIDS imposed at that time. They objected to the long drug approval process, arguing that the risks of experimental interventions were justified for dying patients. This led to programs for granting access to investigational new drugs outside the context of clinical trials, to an expedited review and approval process, and to the use of surrogate endpoints for measuring efficacy (Darrow et al. 2014; FDA 1987, 1990). The current “right to try” movement reflects similar disagreements about how to assess research risks and potential benefits, and how to assign value to risks and potential benefits (Zettler and Greely 2014; Coughlin et al. 2018).

While Mellanby was concerned that excessive caution undermined the interests of society and that volunteers should be allowed to assume risks to serve social interests, he took the safety and welfare of his participants seriously and wanted to ensure that the risks and burdens to which volunteers were exposed were appropriate and justified by the potential benefit to society. We see this when Mellanby describes what happened when he wanted to change the time between scabies infection and treatment. Originally, "volunteers [were told] that they would be infected in various ways (for we were concerned mainly with transmission), but it was understood that as a rule when infection had occurred they would almost immediately be cured" (p. 67). During the study, Mellanby decided that he "wished to follow the course of infections which continued for much longer periods" (p. 67). He explained this to the volunteers, and "they all expressed complete willingness to submit to long periods of infection, although this went far beyond the original experiment" (p. 67). Some of them would have untreated scabies for up to nine months, leaving them with prolonged itching and sometimes resulting in bacterial infections (p. 67). Sometimes, Mellanby notes, he had "to terminate infections [for fear] that permanent damage might be done to the volunteers, though they themselves were quite willing to continue with the experiment" (pp. 67–68). Mellanby demonstrates concern for the risks and burdens participants faced in delaying treatment relative to the knowledge studying prolonged infection might provide.

Whether we approve of his decision to delay treatment for as long as he did or whether we think he struck the right risk-benefit balance is a different matter. Nine months of untreated scabies with no clear reason for needing to observe the course of the disease for *that* long raises serious questions. What did he learn over the extended period of time that he could not have learned in less time? Investigators, sponsors, IRBs, and DSMBs face such challenges today as well. Clinical trials sometimes may be stopped early for safety, efficacy, or commercial reasons (Iltis 2005). Each of these raises different ethical considerations, but all put at risk the ability to learn from a study or to learn as much as could have been learned had the study continued. Sometimes the risk of learning less or even nothing at all is justified because of participant welfare concerns (Brody 1998, p. 156). Studies stopped early because of the observed efficacy of the study intervention, i.e., because the intervention under investigation appears to be working well, sometimes lead to overestimating the benefits of that intervention (Bassler et al. 2010). Overestimating benefits undermines the interests of patients in the future, though failure to stop an investigation early when an intervention truly is beneficial might not only prevent access to an effective therapy but expose participants to unjustified research risks.

Throughout the volume, Mellanby describes his efforts to show respect for his research volunteers' ethical commitments as conscientious objectors. He thought it important that the research always have a civilian application even if it also had a military application, he solicited their input on new lines of research, and he ran the institute as much as possible as the participants saw fit. Although he saw himself as being "in authority….to take all the important decisions regarding the planning of the experiments," he consulted and included participants (p. 37). When Mellanby discusses the idea of adding dietary research, he describes recognizing that "our volunteers being pacifists, we could obviously not work on any question purely of military importance such as the production of emergency rations for mobile troops, but there were many questions regarding the nutrition of the civil population either of Britain or occupied Europe which required elucidation" (p. 51). Perhaps he also did not think he would be successful in getting them to agree to the additional research and thus it was a pragmatic decision, though his tone suggests more than that.

It seems that Mellanby's participants were free to leave at any time, though they did not exercise this option. He updated and consulted with them as his plans changed, and he tried to ensure that no one was permanently harmed by his participation. He notes the importance of "explain[ing] in detail just what was the purpose of anything which we asked them to do" (p. 39). Mellanby notes that he occasionally "conceal[ed] some information, particularly when investigating such subjects as individual reactions to the presence of the causative mites and other topics with a considerable subjective element in them" (p. 40). But, overall, he describes being fairly open with them.

– Despite being attuned to some of the areas we consider in research ethics today, Mellanby's practices were not always consistent with contemporary expectations or norms. For instance, there is a striking difference in how he thought about informed consent documents and the ideal of informed consent described in the literature today. Mellanby did not anticipate permanent harm to volunteers, but he "imagined that it would be a necessary safeguard for each man to sign some sort of detailed contract, setting forth his duties and the risks he was taking, in order to cover [himself] in an emergency and to have something to which to hold a recalcitrant subject if he proved non-cooperative" (p. 36). Interestingly, Mellanby thought that because he was conducting research on a small scale, it was not worth developing such documents. But, he said, "if work of this kind were to be carried out on a large scale and a permanent basis I would agree, though somewhat reluctantly, that the legal side of the matter would have to be looked into very thoroughly, and full scales for compensation and disability would need to be worked out at the outset" (pp. 36–37). Interestingly, he implies that if participants were injured, they would need to be compensated. The United States still does not have a system for compensating people for research-related injuries, which some hold to be deeply problematic. Mellanby saw any document subjects might sign primarily as protecting him and not about sharing information with participants and confirming their voluntary informed consent or protecting participants' interests or rights. Such a document would delineate expectations to ensure that participants knew what they were agreeing to do so that he could remind them of this agreement in the future. Contrast Mellanby's view with contemporary descriptions of the purpose of informed consent for research:

– "Respect for persons requires that subjects, to the degree that they are capable, be given the opportunity to choose what shall or shall not happen to them. This opportunity is provided when adequate standards for informed consent are satisfied" (National Commission for Protection of Human Subjects of Biomedical and Behavioral Research 1979).

- "...the function of the informed consent process is to allow individuals to decide, based on their own preferences and values, whether they want to enroll in a particular study" (Dickert et al. 2017, p. 3).

  "The purpose of informed consent is 2-fold: to ensure that individuals control whether or not they enroll in clinical research and participate only when the research is consistent with their values, interests, and preferences" (Emanuel et al. 2000, p. 2706).

Despite numerous debates regarding informed consent in the literature, it is well-recognized in codes, regulations, and the scholarly literature that informed consent documents are not primarily contracts to enforce future participation or protect researchers and institutions. They are meant to document the voluntary informed agreement of participants. Yet, as Nancy Dubler has put so well, the goal of protecting institutions and researchers seems to drive informed consent documents rather than a desire to respect autonomy by facilitating voluntary informed decision making:

> ... something happened to the doctrine of informed consent on the way to the ball: it got mugged by the corporate, institutional, and administrative risk managers whose focus is singular and directed at the goal of protecting the entity, whatever its form, from possible later liability, not on empowering the patient to make the most individually appropriate decision. This goal – and no other – is reflected in 'informed consent' documents, which neither inform nor empower, but rather dump all of the possibly foreseeable – however remote – risks on the patient (Dubler 2002, pp. 567–8).

Although Mellanby held a radically different understanding of informed consent from the way it is described in the research ethics literature today, much of what passes for informed consent might be driven by interests or concerns similar to his.

What initially appears to be a striking difference between the past and present should instead prompt introspection. Are current policies and practices surrounding informed consent adequate for the ethical conduct of research? If not, where do they fall short and how can those problems be remedied? A wealth of empirical research on informed consent reveals numerous problems. Although some efforts to improve informed consent documents and processes have been implemented, it is likely that more needs to be done to fulfill the stated ideal.

Research and the Role of Institutional Structures

Mellanby mentions several times the importance of both external funding and his independence as a scientist for facilitating progress. He received financial support from the Ministry of the Medical Research Council and describes informing them of his progress. He was grateful that "they had given [him] a completely free hand with the scientific work" and that the experts the ministry consulted about his work "knew that work like this must be run on the spot with a minimum of administrative control" (pp. 69–70). He offers examples of changes he made to his research plan that made his work more useful, such as studying not only the mode of transmission but the course of infections over long periods of time during his investigation (p. 70). Without being able to revise and implement his plans quickly, opportunities to learn would have been lost or significantly delayed. He spoke with the volunteers about the proposed changes and then proceeded rather than having to seek permission from funders or other oversight bodies. This is quite different from what he would be required to do today when conducting research under the auspices of the military, an academic institution, or anyone who required IRB/REC oversight. He attributes much of his progress to the nimbleness afforded him by the lack of administrative control.

Many researchers today lament what they see as impediments to progress that IRBs impose (Schneider 2015; Schneider and Whitney 2011; Rasmussen 2009; Baily et al. 2006; Ledford 2007). The time and energy invested in securing research funding itself is significant (see, for example, Herbert et al. 2013). That effort, coupled with the administrative burden of IRB review and of submitting amendments to the IRB every time an investigator wants to modify a research plan, involve a substantial investment of time. In some cases, the time required might serve as a disincentive for investigators to improve their studies. The investment in thoroughly vetting detailed research proposals which, like preparing grant applications, consumes significant human capital, does not guarantee that funded projects will make significant contributions. Nor does the careful review of human research proposals by institutional review boards/research ethics committees (IRBs/RECs) and the other regulatory and oversight mechanisms guarantee that all research will be conducted ethically. However, given the history of human research abuses, leaving scientists independently to determine what counts as good science or science worth doing and how humans may be used might not be the answer to these burdens. Science, after all, is a social endeavor and requires social trust (Robert 2009; Hurlbut and Robert 2012; London 2012). Yet, Mellanby's comments, coupled with contemporary criticisms of various research processes, prompt us to ask about improvements that could be made to advance the goals of ensuring that scientists earn and maintain trust, use resources wisely, and avoid violating the rights and interests of human beings while at the same time promoting research to improve health and well-being. Insofar as unnecessary or poorly implemented oversight plans undermine the value or validity of human research without a compensating benefit, such as protecting participants' welfare or rights, those policies and practices raise ethical concerns.

In His Own Voice: First Person Narratives in Bioethics

Personal stories yield insights into ethical decision-making and can make us aware of ethical issues or considerations that we otherwise not identify. If we read only Mellanby's research publications, such as his 1941 *British Medical Journal* publication, we would know this about his scabies research:

> These experiments were made using volunteers who lived under controlled institutional conditions and who were subjected to different types of contact with scabies infection. The volunteers were all pacifists who had offered to cooperate in this work and who, during the experiments, received board and lodging and a weekly payment similar to that which they would have received had they been called up for military service. They agreed to submit themselves to infection and to allow the course of the disease to be followed on their persons (1941, p. 405).

Through *Human Guinea Pigs*, we gain a better understanding of the burdens research participation imposed. It's one thing to say that "the course of the disease [was] followed on the person." It is quite another thing to read this:

> Certain volunteers were reduced to sleeping naked as they scratched so vigorously in their sleep that their pyjamas were torn to shreds. Unfortunately, this constant scratching not infrequently damaged the skin and allowed the entry of sepsis-producing bacteria; these gave rise to unpleasant secondary infection, which could not be cleared up until the scabies was cured (p. 67).

The description in *Human Guinea Pigs* is much more vivid and is more likely to make us question whether the prolonged period of infection without treatment was justified. Hearing from participants themselves would have furthered our understanding of how they viewed their participation and whether they believed that they had a voice in shaping the research, felt free to leave the study, saw Mellanby's treatment of them as fair, and understood what they were agreeing to do when they said yes, for example.

It might be especially interesting to know what his participants thought about being turned into investigators themselves who studied scabies and the prevention of scabies in school children. In *Human Guinea Pigs*, Mellanby explains that his "volunteers did a lot of useful 'field work' connected with the scabies investigation" (p. 75). This included traveling to schools to examine children for scabies, to apply possible prophylactic "treatments" to the children, and to follow the prevalence of scabies among the children. A search of the medical literature reveals a 1944 publication by Mellanby in the *British Medical Journal* reporting on this scabies research in school children that corresponds to the work he describes in his book. One would never know from reading this *BMJ* paper that most of the work was carried out by the conscientious objectors who themselves were Mellanby's research subjects. The range of roles his volunteers played and the fact that they were themselves engaged as researchers raises numerous ethical questions as does the research on children. Unfortunately, Mellanby does not elaborate sufficiently in *Human Guinea Pigs* to know, for example, whether the parents had any role in deciding what would be done to their children. Nevertheless, were it not for the book, we would not even know that the subjects had been used to conduct research on children because this interesting and important detail is missing from the medical report of the investigation. Further consideration of the use of participants as researchers might prompt introspection about community based participatory research practices today.

Mellanby's narrative approach paints a more nuanced picture of his ethical judgment and exposes the reasons he made certain decisions and the factors he considered in conducting his work. The evidence of ethical reasoning in *Human Guinea*

Pigs, even where we find flaws, reveals facts of the history of research and research ethics that we otherwise might have missed. Ethical considerations that some scientists today might find burdensome were not necessarily invented by outsiders. Rather, they are obligations that their predecessors had envisioned. After all, Mellanby implicitly or explicitly mentioned many obligations routinely cited today as central to the ethical conduct of human research. Making available Mellanby's first person account through this re-printing also gives the commentary authors the opportunity to draw out themes and lessons that might have been lost or gone unnoticed if we only had access to Mellanby's scientific publications. Inquiring into the past can help us interrogate current practices. Sometimes we might find similar patterns of reasoning or behavior that disturb us. Where we thought we had come a long way, we find that we have not. This should motivate a more careful assessment of the present and a better orientation for the future.

References

Ackerman, Terrence F. 1989. An ethical framework for the practice of paying research subjects. *IRB: Ethics & Human Research* 11 (4): 1–4.

Appelbaum, Paul S., Charles W. Lidz, and Robert Klitzman. 2009. Voluntariness of consent to research: A conceptual model. *Hastings Center Report* 39 (1): 30–39.

Avorn, J. 2005. FDA standards—Good enough for government work? *New England Journal of Medicine* 353 (10): 969–972.

Baily, Mary Ann, et al. 2006. Special report: The ethics of using QI methods to improve health care quality and safety. *Hastings Center Report* 36 (4): S1–S40.

Baker, Robert. 2019. Human Guinea-Pigs: Mellanby, Pappworth and Club Reguation. In *TITLE*, ed. L. Rasmussen. Dordrecht: Springer.

Bassler, D., M. Briel, V.M. Montori, M. Lane, P. Glasziou, Q. Zhou, et al. 2010. Stopping randomized trials early for benefit and estimation of treatment effects: Systematic review and meta-regression analysis. *JAMA* 303 (12): 1180–1187.

Bennett, J. Claude. 1993. Inclusion of women in clinical trials--Policies for population subgroups. *New England Journal of Medicine* 329 (4): 288–292.

Berger, Jeffrey S., Chiara Melloni, Tracy Y. Wang, Rowena J. Dolor, Camille G. Frazier, Zainab Samad, Eric D. Peterson, Daniel B. Mark, and L. Kristin Newby. 2009. Reporting and representation of race/ethnicity in published randomized trials. *American Heart Journal* 158 (5): 742–747.

Brody, B. 1998. *The ethics of biomedical research*. New York: Oxford University Press.

———. 2002. Ethical issues in clinical trials in developing countries. *Statistics in Medicine* 21 (10): 2853–2858.

Christopher, P.P., P.J. Candilis, J.D. Rich, and C.W. Lidz. 2011. An empirical ethics agenda for psychiatric research involving prisoners. *AJOB Primary Research* 2 (4): 18–25.

Corbie-Smith, Giselle, D.M. St George, Sandra Moody-Ayers, and David F. Ransohoff. 2003. Adequacy of reporting race/ethnicity in clinical trials in areas of health disparities. *Journal of Clinical Epidemiology* 56 (5): 416–420.

Corbie-Smith, Giselle, Sandra Moody-Ayers, and Angela D. Thrasher. 2004. Closing the circle between minority inclusion in research and health disparities. *Archives of Internal Medicine* 164 (13): 1362–1364.

Coughlin, C.N., N.M. King, and M. McKinney. 2018. Regenerative medicine and the right to try. *Wake Forest Journal of Business and Intellectual Property Law* 18 (4): 592–637.

Council for International Organizations of Medical Sciences. 2002. International ethical guidelines for biomedical research involving human subjects. *Bulletin of Medical Ethics* 182: 17.
Darrow, J.J., J.J. Avorn, and A.S. Kesselheim. 2014. New FDA breakthrough-drug category – Implications for patients. *New England Journal of Medicine* 371 (13): 1252–1258.
Dickert, Neal, and Christine Grady. 1999. What's the price of a research subject? Approaches to payment for research participation. *New England Journal of Medicine* 341 (3): 198–203.
Dickert, Neal, Ezekiel Emanuel, and Christine Grady. 2002. Paying research subjects: An analysis of current policies. *Annals of Internal Medicine* 136 (5): 368–373.
Dickert, Neal W., Nir Eyal, Sara F. Goldkind, Christine Grady, Steven Joffe, Bernard Lo, Franklin G. Miller, Rebecca D. Pentz, Robert Silbergleit, Kevin P. Weinfurt, David Wendler, Scott Y. H. Kim. 2017. Reframing consent for clinical research: A function-based approach. *The American Journal of Bioethics* 17 (12): 3–11.
Dresser, Rebecca. 1992. Wanted: Single, white male for medical research. *Hastings Center Report* 22 (1): 24–29.
Dubler, N. 2002. New directions in human subject research: Looking beyond the academic medical center: Remaining faithful to the promises given: Maintaining standards in changing times. *Seton Hall Law Review* 32 (563): 563–571.
Edens, J.F., M. Epstein, P.G. Stiles, and N.G. Poythress Jr. 2011. Voluntary consent in correctional settings: Do offenders feel coerced to participate in research? *Behavioral Sciences & the Law* 29 (6): 771–795.
Emanuel, Ezekiel J. 2004. Ending concerns about undue inducement. *The Journal of Law Medicine & Ethics* 32 (1): 100–105.
Emanuel, E.J., D. Wendler, and C. Grady. 2000. What makes clinical research ethical?. *Journal of the American Medical Association* 283 (20): 2701–2711.
FDA. 1987. Investigational new drug, antibiotic and biological drug product regulations: Treatment use and sale. *Federal Register* 52: 19476.
———. 1990. Expanded availability of investigational new drugs through a parallel track mechanism for people with AIDS and HIV-related disease. *Federal Register* 55: 20857.
Fisher, Jill A. 2013. Expanding the frame of 'voluntariness' in informed consent: Structural coercion and the power of social and economic context. *Kennedy Institute of Ethics Journal* 23 (4): 355–379.
Fisher, Jill A., Lisa McManus, Megan M. Wood, Marci D. Cottingham, Julianne M. Kalbaugh, Torin Monahan, Rebecca L. Walker. 2018. Healthy volunteers' perceptions of the benefits of their participation in Phase I clinical trials. *Journal of Empirical Research on Human Research Ethics* 13 (5): 494–510.
Glasziou, P., D.G. Altman, P. Bossuyt, I. Boutron, M. Clarke, S. Julious, et al. 2014. Reducing waste from incomplete or unusable reports of biomedical research. *The Lancet* 383 (9913): 267–276.
Grady, Christine. 2001. Money for research participation: Does it jeopardize informed consent? *American Journal of Bioethics* 1 (2): 40–44.
Halpern, Scott D., Jason H.T. Karlawish, and Jesse A. Berlin. 2002. The continuing unethical conduct of underpowered clinical trials. *JAMA* 288 (3): 358–362.
Herbert, Danielle L., Adrian G. Barnett, Philip Clarke, and Nicholas Graves. 2013. On the time spent preparing grant proposals: An observational study of Australian researchers. *British Medical Journal Open* 3 (5): e002800.
Hoel, Andrew W., Ahmed Kayssi, Soma Brahmanandam, Michael Belkin, Michael S. Conte, and Louis L. Nguyen. 2009. Under-representation of women and ethnic minorities in vascular surgery randomized controlled trials. *Journal of Vascular Surgery* 50 (2): 349–354.
Hurlbut, J. Benjamin, and Jason Scott Robert. 2012. Stem cells, science, and public reasoning. *Journal of Policy Analysis and Management* 31 (3): 707–714.
Iltis, A.S. 2005. Stopping trials early for commercial reasons: The risk–benefit relationship as a moral compass. *Journal of Medical Ethics* 31 (7): 410–414.

Iltis, Ana S. 2016. Prenatal screening and prenatal diagnosis: Contemporary practices in light of the past. *Journal of Medical Ethics* 42: 334–339.
International Conference on Harmonisation. 1996. *ICH topic E6. Guideline for good clinical practice*. Geneva: International Conference on Harmonisation.
Ioannidis, John P.A. 2005. Why most published research findings are false. *PLoS Medicine* 2 (8): e124.
———. 2014. Clinical trials: What a waste. *BMJ: British Medical Journal (Online)* 349: g7089.
———. 2015. How to make more published research true. *Revista Cubana de Información en Ciencias de la Salud (ACIMED)* 26 (2): 187–200.
———. 2016. Why most clinical research is not useful. *PLoS Medicine* 13 (6): e1002049.
King, N.M.P., and L. Churchill. 2008. Assessing and comparing potential benefits and risks of harm. In *The Oxford textbook of clinical research ethics*, ed. E. Emanuel, C. Grady, R. Crouch, et al., 514–526. New York: Oxford University Press.
Kopelman, Loretta M. 2000. Moral problems in assessing research risk. *IRB: Ethics & Human Research* 22 (5): 3–6.
LaFleur, William, Gernot Böhme, and Susumu Shimazono, eds. 2007. *Dark medicine: Rationalizing unethical research*. Bloomington: Indiana University Press.
Ledford, Heidi. 2007. Human-subjects research: Trial and error. *Nature* 448 (7153): 530–532.
London, A.J. 2012. A non-paternalistic model of research ethics and oversight: Assessing the benefits of prospective review. *The Journal of Law, Medicine & Ethics* 40 (4): 930–944.
Macklin, Ruth. 1981. On paying money to research subjects: 'Due' and 'undue' inducements. *IRB: A Review of Human Subjects Research* 3: 1–6.
Mazure, Carolyn M., and Daniel P. Jones. 2015. Twenty years and still counting: Including women as participants and studying sex and gender in biomedical research. *BMC Women's Health* 15 (1): 94.
McCabe, L.L., and E.R. McCabe. 2011. Are we entering a "perfect storm" for a resurgence of eugenics? Science, medicine, and their social context. In *A century of eugenics in America: From the Indiana experiment to the human genome era*, ed. P.A. Lombardo, 193–218. Bloomington: Indiana University Press.
McNeil, P. 1997. Paying people to participate in research: Why not? A response to Wilkinson and Moore. *Bioethics* 11: 390–396.
Mehlman, M.J. 2011. Modern eugenics and the law. In *A century of eugenics in America: From the Indiana experiment to the human genome era*, ed. P.A. Lombardo, 219–240. Bloomington: Indiana University Press.
Mellanby, K. 1941. The transmission of scabies. *British Medical Journal* 2 (4211): 404–406.
———. 1944. Experiments on scabies prophylaxis. *British Medical Journal* 1 (4350): 689–690.
Miller, Franklin G., and Stephen Joffe. 2009. Limits to research risks. *Journal of Medical Ethics* 35 (7): 445–449.
Moser, D.J., S. Arndt, J.E. Kanz, M.L. Benjamin, J.D. Bayless, R.L. Reese, et al. 2004. Coercion and informed consent in research involving prisoners. *Comprehensive Psychiatry* 45 (1): 1–9.
National Commission for Protection of Human Subjects of Biomedical and Behavioral Research. 1979. *Belmont Report: Ethical Principles and Guidelines for the Protection of Human Subjects of Research*. Department of Health, Education, and Welfare.
Nelson, Robert M., and Jon F. Merz. 2002. Voluntariness of consent for research: An empirical and conceptual review. *Medical Care* 40: V69–V80.
Nuremberg Code. 1949. *Trials of war criminals before the Nuremberg Military Tribunals under Control Council Law no. 10*. Vol. 2, 181–182. Washington, DC: U.S. Government Printing Office. Available online: https://history.nih.gov/research/downloads/nuremberg.pdf. Accessed 10 Dec 2018.
Rasmussen, Lisa M. 2009. Problems with minimal-risk research oversight: A threat to academic freedom? *IRB: Ethics & Human Research* 31 (3): 11–16.
Reverby, S.M. 2011. 'Normal exposure' and inoculation syphilis: A PHS 'Tuskegee' doctor in Guatemala, 1946–1948. *Journal of Policy History* 23: 6–28.

Reverby, Susan M. 2012. Ethical failures and history lessons: The US Public Health Service research studies in Tuskegee and Guatemala. *Public Health Reviews* 34 (1): 13.
Robert, Jason Scott. 2009. Toward a better bioethics. *Science and Engineering Ethics* 15: 283–291.
Ross, Joseph S., Marian Mocanu, Julianna F. Lampropulos, Tony Tse, and Harlan M. Krumholz. 2013. Time to publication among completed clinical trials. *JAMA Internal Medicine* 173 (9): 825–828.
Schneider, Carl. 2015. *The Censor's hand: The misregulation of human-subject research.* Cambridge, MA: MIT Press.
Schneider, Carl, and Simon, Whitney. 2011. Rethinking the IRB's role. *Presentation at Wake Forest University*, November 11.
Shahar, Eyal. 2003. Generalizability: Beyond plausibility and handwaving. *Journal of Evaluation in Clinical Practice* 9 (2): 151–159.
Stiles, Paul G., Monica Epstein, Norman Poythress, and John F. Edens. 2012. Protecting people who decline to participate in research: An example from a prison setting. *IRB-Ethics and Human Research* 34 (2): 15.
Weijer, Charles, and Paul B. Miller. 2004. When are research risks reasonable in relation to anticipated benefits? *Nature Medicine* 10 (6): 570.
Wendler, David, and Franklin G. Miller. 2007. Assessing research risks systematically: The net risks test. *Journal of Medical Ethics* 33 (8): 481–486.
World Medical Association. 2013. World Medical Association Declaration of Helsinki: Ethical principles for medical research involving human subjects. *JAMA* 310 (20): 2191.
Zettler, P.J., and H.T. Greely. 2014. The strange allure of state 'right to try' laws. *JAMA Internal Medicine* 174 (12): 1885–1886.

Chapter 20
The 'Untrammelled' Scientist and the 'Normal' Volunteer: Some Reflections

Alastair V. Campbell

These reflections on Kenneth Mellanby's idiosyncratic approach to research with human subjects have been provoked by a number of statements in the first (1945) edition of *Human Guinea Pigs*, some of which are amplified in the second (1973) edition. In the Preface to the first edition, Mellanby writes: "Some scientists feel there is no longer a place for the individual carrying out research in his own particular way…but I personally also think that scientific progress will be sorely stunted if there is no place for the individual to develop his own particular type of work" (Mellanby 1973, 7). He continues in the same vein in the next paragraph: "I hope there will always be a place, and funds, for the individual who wishes to work in his own way, untrammelled as much as possible by the "red-tape" which seems to be a necessary accompaniment of any large-scale organisation" (id.loc.).

This approach to independent governance might be named, perhaps rather unkindly, the 'lone ranger' style of medical research, and possibly it was not uncommon at the time when Mellanby was carrying out his research into scabies, and subsequently into other conditions. However, it should not be lightly dismissed as merely amusing, quirky, or a product of the times. Rather, it raises some very critical issues (as subsequent events were to show) about the risks of allowing 'untrammelled' scientists to decide for themselves the design of research with human subjects, including how subjects should be recruited, what methods should be used, and how risks and benefits should be assessed. In the first sections of this chapter I shall consider these issues in some detail, and I shall also analyse the claim that it is unfair to judge research carried out in the 1940s by standards laid down several decades later.

A second feature of Mellanby's approach is his use of conscientious objectors as research subjects – 'human guinea pigs', as he himself describes them. One striking

A. V. Campbell (✉)
National University of Singapore, Singapore, Singapore
e-mail: alastairandsally@gmail.com

L. M. Rasmussen (ed.), *Human Guinea Pigs, by Kenneth Mellanby: A Reprint with Commentaries*, Philosophy and Medicine 134,
https://doi.org/10.1007/978-3-030-37697-0_22

feature of his account of this aspect of his research is his ambivalent attitude to these volunteers, summed up by the title of the chapter in which he discusses their personal characteristics in some detail – 'The *Problem* of the Conscientious Objector' (emphasis added). Clearly, he is puzzled by what motivates people (in particular, men) to refuse to fight in war. He himself was keen to sign up from the start of WW2, but prevented from doing so at first, since, as a biologist, he was regarded as being in a reserved occupation. However, eventually he was allowed to join the army, and he served in North Africa, South-East Asia and New Guinea. Although he writes in places in very positive terms about his volunteers, and mentions that they seemed to be fully accepted by the military volunteers he used later, he struggles to understand how they can be regarded as 'normal'. Thus, he describes his laboratory assistant, Walter Bartley, a conscientious objector, as "an efficient worker and otherwise normal person" (Mellanby 1973, 34); and in similar language, he says of his first two volunteers, "They seemed to be more or less normal people" (id. loc.). We might be left wondering how Mellanby defines the term 'normal', but since these first volunteers are clearly middle class (a maths teacher and an artist), it seems likely that 'normal' would be best translated as "people like us, good chaps basically!"

At this point I should mention that I have a personal interest in this aspect of Mellanby's research, since my father was a conscientious objector in WWI.[1] This ambivalence in Mellanby's attitude to his research subjects raises a more general question about the voluntariness of participation in research by conscientious objectors, especially in wartime. Moreover, since all matters of consent were handled solely by Mellanby, we must wonder just how his ambivalent and puzzled attitude affected the manner in which specific consent was obtained, especially for the more risky and unpleasant aspects of the project.

Research Governance and the Lone Ranger

We saw above how Mellanby's resistance to what he sees as unwarranted interference by committees in his research endeavours was strongly expressed in the Preface to the first edition. Astonishingly, we find this even more forcefully expressed in the second edition published in 1973. I am quoting from the final paragraph of this second preface at some length, since it shows what a cavalier attitude the author has to questions of risk in medical research:

> There are clearly still many gaps in medical knowledge. Progress in chemotherapy is being delayed because, in the hope of making sure that every drug is completely safe, almost impossibly stringent standards are being insisted upon. I think we still need facilities by which volunteers anxious to promote medical progress can be allowed to give their ser-

[1] I shall be referring to my father's thoughts and motivations later, when I look in more detail at the participation of conscientious objectors as volunteers in research.

> vices, even if there is some degree of risk to their own health, or even, in extreme cases, to their own lives (Mellanby 1973, 12).

Quite apart from any indication of what Mellanby's conception of "almost impossibly stringent standards" is, or of what 'extreme cases' he had in mind when recommending that volunteers being invited to risk their lives for the future of mankind, the most astonishing aspect of this polemic against safety standards in research is that it was written more than a decade after the Thalidomide tragedy and nearly ten years after the World Medical Association had published the first version of the Helsinki Declaration. Clearly nothing has changed in the mind of the author, and responsible research governance is still seen as an anathema to this self-styled scientific pioneer, who claims to have directed "one of the most successful programmes ever initiated" (Mellanby 1973, 9), untrammelled, of course, by government interference.

In fact, as noted elsewhere in this volume (see Chapter "*Human guinea-pigs*: Mellanby, Pappworth and Club Regulation" by Baker), the main stimulus for this second edition of the book seems to have been Mellanby's annoyance that his title had been 'plagiarised' by Pappworth's *Human Guinea Pigs: experimentation on man*, published in 1967. This volume, like H. K. Beecher's book on the same topic (Beecher 1970), revealed in great detail how many unethical experiments were still being carried out by the medical profession despite the requirements of the Nuremberg Code. Mellanby dismisses Pappworth's concerns as "expressed somewhat emotionally" (1973, 11), and complains that the book does not mention *his* research at all.[2] What we see here, then, is Mellanby's entirely negative view of any kind of external review of medical research, believing it to be an unwarranted and detrimental interference in the autonomy of the medical researcher, and, moreover, an unjustified querying of the humanitarian concerns of a noble profession. This view of ethical appraisal of medical research was clearly still prevalent in the 1960s and 1970s, and Mellanby was by no means alone among his medical and scientific colleagues in holding such negative views. This is clear from Pappworth's experience, described in the Preface to his book:

> During private discussion of this subject I have been frequently attacked by doctors who contend that by such publication I am doing a great disservice to my profession, that I am undermining the faith and trust that lay people have in doctors...I have been subjected to frequent telephone calls, almost entirely from strangers, in an attempt to persuade me to abandon the project (Pappworth 1967, x).

Clearly, then, Mellanby's strictures against external review of his work provide considerable insight into how a lack of awareness among medical and scientific researchers of several key ethical concerns in research could lead to some of the research scandals that were to come to light later. As his work was of relatively low risk (though not entirely risk-free), we can look at its deficiencies in research governance in a paradigmatic way, without necessarily denigrating its beneficial

[2] This is somewhat ironic, as Pappworth was listing only those projects which he considered to be unethical!

outcomes. In other words, we can use it to show the dangers of the 'lone ranger' approach, even if in this case these dangers seem to have been partly avoided. Later in this section, I shall look at a much more serious example of allowing a researcher to act 'untrammelled' by effective committee oversight (cervical cancer research carried out by a single researcher in New Zealand in the 1960s), which had major detrimental consequences for a large group of women. Discussing this more extreme example will also enable me to tackle the question of whether applying current research standards to past research is anachronistic, and therefore unfair.

Central Ethical Issues in Biomedical Research

There are three aspects of biomedical research in which good ethical governance is essential: risk and benefit assessment; information and consent; and voluntariness, including the right to withdraw. I shall consider each of these as they apply to Mellanby's main research project, which was establishing the most effective methods for treating and preventing the transmission of scabies.

Risk and Benefit

I shall not attempt to give a full account of Mellanby's scabies project. It appears to have been a prolonged and constantly evolving series of experiments with volunteers in a communal residential setting, which was under constant management by one individual, who was also the principal, and usually the sole, investigator. Given the highly diffuse nature of Mellanby's anecdotal account of the main project and sub-projects and the absence of any clear documentation of the results, it is very hard to pin down just exactly what the research participants were subjected to at any given time, how long some of the experimental interventions continued, and what criteria were used to discontinue or change the various procedures used. In short, apart from one brief account of the proposal prepared for the Ministry of Health in 1940 (Mellanby 1973, 21–26), we have absolutely no research protocols on which to base an assessment of the various risks and benefits. What we can deduce is that a project originally planned to last for one year, lasted for at least three and a half years, and that the original agreement with the volunteers was that, once infected with scabies, they would "almost immediately be cured" (67). However, later this was entirely changed to a project that was studying the natural history of the disease for nine months or more. Clearly the whole risk aspect of the project was changed dramatically by this alteration in research design. Mellanby's own description of what the volunteers endured in this later phase of the experiment vividly conveys how risky it was:

> This experiment was very unpleasant for the participants…Some kept rough brushes to rub over the skin to relieve irritation. On cold nights some would rise from a sleepless bed and walk naked through the house…Certain volunteers were reduced to sleeping naked as they scratched so vigorously that their pyjamas were torn to shreds. Unfortunately, this constant scratching not infrequently damaged the skin and allowed the entry of sepsis-producing bacteria; these gave rise to unpleasant secondary infection, which could not be cleared up until the scabies was cured. I had sometimes to terminate infections as I was afraid that permanent damage might be done to volunteers (id loc.).

What then of risk-benefit analysis? Clearly scabies was a fairly serious public health problem at the time, though it was not in itself a long-term danger to health or survival. So, it could be argued that there was benefit to the general public in finding effective ways of both preventing its transmission and curing the individuals who suffered from it as quickly as possible and with the least uncomfortable treatment methods. Moreover, since scabies was fairly widespread in the Armed Forces, these benefits were of considerable interest to the military authorities, especially in wartime. It is also obvious that there was no benefit at all to the volunteers (apart from some factors to do with their wish to be seen as doing worthwhile and courageous work, which I shall consider later). Does this mean that the risks to the volunteers were proportional and justified? We learn from Mellanby himself that this question was raised in Parliament. In reply to a question from a Labour MP, the Minister of Health gave this reply regarding "whether there were any temporary or permanent dangers attaching to it": "[The volunteers] have suffered discomfort rather than danger, but I am advised that they have not been free from risk of some prolonged disability" (Mellanby 1973, 69). One looks in vain for any clarification by Mellanby of this Parliamentary answer. Advised by whom, one wonders? And what prolonged disability was a possible risk? Indeed, Mellanby's casual and subjective treatment of risk is one of the main ethical deficiencies in this whole project. As the quotation from the book about the later phase illustrates, *he* was the sole judge of whether to initiate treatment of a septic infection, and he appears to have done this reluctantly and in the absence of any specified criteria for assessing the degree of risk. Moreover, he was a biologist, not a medical practitioner, who was making judgment calls about an infection which he describes at first as merely 'unpleasant' rather than possibly causing 'permanent damage', as he acknowledges in the next sentence.

We have here a clear illustration of why a full protocol detailing in advance all the proposed research methods and their associated risks, plus the criteria for discontinuation or modification of the research, is absolutely essential. This protocol must be fully assessed by an independent ethics committee, who can reject the proposal or require minor or major modifications. This is not 'red tape', but an essential protection for all research participants, whose welfare should not be subject to judgment calls by the researcher, for whom there is an obvious motivation to complete the research. Such disregard for objective standards of risk assessment cannot be justified, as Mellanby seems to believe, by claiming that the participants were willing to take the risks (a point to which I shall return later in this chapter).

Consent

The claim that the volunteers were happy to undergo risky procedures takes us to the second topic: the requirement for informed consent. This is one of the principal requirements of the Nuremberg Code, and it was re-enforced and more fully elaborated in the Helsinki Declaration. One of the main issues determining the validity of consent is the quality of the information given to participants, and, for this reason, research ethics committees pay very close attention to both the information sheets provided by researchers and the wording of the consent forms to be signed by participants. We have no evidence that Mellanby provided any such written documents to his volunteers. On the contrary, he explicitly rejects having them sign "some sort of detailed contract, setting forth his duties and the risks he was taking." This he regards as too complicated, costly, and legalistic, and instead he claims, "I think much of the success of the experiment was due to its air of informality" (1973, 36). In the absence of any such documents, we have to rely entirely on Mellanby's reassurances that the volunteers were fully informed of all that would be entailed in joining this research project, including the various methods that would be employed and the risks of discomfort, loss of sleep, septic infection and possible adverse consequences on their future health.

We see, then, that the question of whether there was valid and adequate consent is a very serious one for this research project. Obviously, we should not have to guess about what information the volunteers were actually given at enrollment and at the later stages of the research. All we have are a few clues about how he shared information. He says at one point that in "the managing of volunteers" it was essential "to be willing to take the trouble to explain in detail just what was the purpose of anything which we asked them to do" (38). The wording is significant here. There is no suggestion that the volunteers have the *right* to this information, but rather one must take this *trouble* in order to manage them effectively. Periodical meetings were held to give a progress report and answer any questions, albeit these questions from the participants were "often not very profound" (id. loc.)! However, not all information was shared: "As a rule, I took volunteers entirely into my confidence about the progress of the work, though on occasion it was necessary to conceal some information" (id. loc.). Thus, Mellanby sees himself as the sole judge of this concealment of information, and he repeats his attack on control by a committee, concluding that "this is the one field where a Dictator is required!" (39). We have simply no clue about whether the concealment of information was justified, or even whether the participants were given the information later in a de-briefing.

As I remarked earlier, one might be tempted to dismiss these extraordinary assertions of justified 'autocracy' in the conduct of research as merely quirky, and explicable in terms of Mellanby's apparently judgmental and self-confident personality. However, the issues are too serious to be minimised in this way. In this 'lone ranger' approach to research governance, the way is wide open to the total disrespect and serious exploitation of research subjects through the 'missionary zeal' (Mellanby's own phrase) of the researcher. As we shall see later, such an attitude can lead to even more serious harm to research participants than the frequently unpleasant and occasionally dangerous interventions that Mellanby was able to introduce, without any scrutiny or control by his much-hated committees.

Voluntariness

Later this chapter I shall look in more detail at the specific issues arising from the use of conscientious objectors as volunteers in biomedical research. At this stage I shall consider only the broader issues of how volunteering is to be understood and how genuine voluntarism is to be ensured. We can define volunteering as giving fully informed consent to participate in research in the absence of coercion or undue influence. As so often in such definitions, the devil is in the detail. When is influence 'undue'? And what is the range of coercion beyond the obvious examples of the use of force or the threat of physical harm? Discussions of research ethics over the 70 years since the Nazi experiments on the helpless inmates of death camps have looked at the coercion or influence created by situations of dependency, such the doctor-patient relationship, the student-teacher relationship, and the limited choices of prison inmates. In addition, economic dependency has been explored in relation to the amount of monetary payment offered to research volunteers. Finally, the general question of types of vulnerability has raised the possibility that whole groups of potential research subjects, such as pregnant women, minors, prisoners, and victims of natural disasters, should either be excluded from participation or be restricted to research projects with minimal risk (ten Have 2016).

The unusual setting in which Mellanby's research was conducted does raise questions about the voluntariness of the continued participation of his research participants. He created a residential community in which household tasks were shared and the common goal of fostering his research was re-enforced in community meetings. This group also shared the burdens and potential embarrassment of the various interventions, including wearing the underclothing of infected soldiers, sleeping naked under dirty bedclothes, which may have been infested with the scabies mites, having mites from infected persons physically introduced into their skin, and sharing a bed with an infected colleague. Added to this, despite their differences in many ways, the participants all shared a moral commitment to pacifism. Under these conditions, it must have been hard for individuals to refuse to undergo some of the procedures, or to withdraw from the research altogether (which would presumably entail leaving the community). In this context we can reasonably question Mellanby's confident assertion that, even in the risky situation of secondary infection, his volunteers were willing to continue. Furthermore, it remains unclear how many participants withdrew from the research, and we are not given any account of their reasons for doing so. All that is provided is this vague statement: "The membership of the household changed a little from time to time, though after three-and-a-half years some of the original volunteers were still in residence and taking part in experiments" (1973, 65).

As with the rest of this very anecdotal account of the research, we are left with no clear picture of how voluntary participation in the continuing research was ensured, how easy or difficult it was for individuals to opt out partly or entirely, and how many of them did when some of the details of the experiments became clear, and what pressure, moral or financial, might have made it difficult for some participants to leave the community. Unfortunately, Mellanby seems more interested in explaining the domestic arrangements, such as cooking and house cleaning, than in offering any genuine insight into how the voluntariness of the research was ensured.

The Unfortunate Experiment

In the final parts of this section of the chapter I shall demonstrate why the ethical issues raised by Mellanby's research are of such vital importance, by referring to another example of 'lone ranger' research, which had devastating consequences for the unfortunate women caught up in it. The example I am using is the so-called 'unfortunate experiment' in Auckland Women's Hospital in New Zealand, which led eventually to a Judicial Inquiry, at which I was an expert witness (and subsequently to a Disciplinary Hearing by the NZ Medical Council, at which I was also called as a witness). There is a long and very detailed Report of the Inquiry (Cartwright 1988), but here I shall give just some of the basic issues, which illustrate the serious dangers of absent, or inadequate, external ethical appraisal of research projects.

The term 'unfortunate experiment' was coined in a letter in the *NZ Medical Journal* written by Professor David Skegg, Professor of Preventive and Social Medicine at Otago Medical School. Skegg was responding to a letter from Associate Professor Herbert Green, which had opposed the introduction of a national cervical screening programme in New Zealand, on the grounds that it was unnecessary. Skegg wrote: "The case for the effectiveness of screening does not rest on the unfortunate experiment at National Women's hospital, in which women with abnormal smears were treated conservatively and a proportion have developed invasive cancer" (quoted in Manning 2009, 27).

Skegg wrote this letter in 1986, but the research by Green to which he is referring began 20 years earlier, in 1966. Concern about the project had been expressed over several years by a number of Green's colleagues, and articles documenting these concerns had been published in medical journals, but nothing was done about stopping the project until an article was published in *Metro*, a popular monthly magazine, in June 1987, exposing what had been happening to women over two decades. The uproar this article caused led the Minister of Health to set up a judicial inquiry, which became known as the *Cartwright Inquiry*, after the judge appointed to head it, Silvia Cartwright.

What, then, was this 'unfortunate experiment'? Green, who was a consultant in Obstetrics and Gynaecology in the Postgraduate Medical School at Auckland Women's Hospital, had come to the belief that carcinoma in situ (CIS), a lesion of the surface of the cervix, which could be detected by a cervical smear test (PAP smear), was not a precursor to invasive cancer. Standard practice in the 1960s, following a positive result from a PAP smear, was to carry out a cone biopsy. This entailed removing a cone shaped section of tissue from the cervix in the area which contained the cancerous cells, so, it was hoped, preventing the spread of cancer. This procedure was much preferred to earlier interventions, which usually entailed a hysterectomy, with its associated bad side effects including loss of fertility. Contrary to received medical opinion, both nationally and internationally, Green hypothesised that CIS was a benign condition and that it could be safely treated 'conservatively', even after a positive smear test result. To prove this hypothesis, he took a proposal to a meeting of the hospital's senior medical staff. The Report of the Cartwright Inquiry describes this event as follows:

> Herbert Green laid before the meeting a written proposal in which he argued that 'the time had come' for all women patients under the age of thirty-five with positive smears (CIS and no clinical or colposcopic evidence of invasive cancer) to be diagnosed and treated by 'lesser procedures'. This meant they would not have a hysterectomy or cone biopsy, but would be observed and punch specimens would be taken for diagnostic purposes. It was proposed that, 'in the interests of continuity of supervision and patient-confidence', all such patients should be referred to Green, 'whose conscience is clear and who could therefore accept complete responsibility for whatever happens' (Manning 2009, 29).

The closing phrase is significant, since it shows that from the outset Green was aware that his proposed study was risky, and yet over two decades he flatly refused to accept the increasing evidence from highly qualified colleagues (a colposcopist and a pathologist) that, on the basis of the samples he was providing, his hypothesis was wrong. Green either disputed the accuracy of his colleagues' findings or suggested that in the case of those patients with invasive carcinoma, it had already been present, though undetected, prior to them being included in his research group. Thus, the outcome for the patients enrolled in this study was devastating: not only did they have repeated visits for 'follow-up' over very many years, but, despite all of this supposed monitoring, many of them then had to endure operations for much more advanced cancer than would have been necessary with early detection. One study, using the data from Green's own study, showed that patients with continuing abnormal smears were 24.8 times more likely to develop invasive cancer than patients with normal smears two years after diagnosis of CIS and with standard treatment given.

There are a number of features relevant to research governance in this sorry tale of medical malpractice.[3] The following are the most pressing areas of ethical concern:

Firstly, there was *no consent* for this research project. At no time were the women in Green's study told that they were part of a research project or that what they were receiving was unusual and potentially highly risky management of their condition.

[3] Green's head of department, Dennis Bonham, was found guilty of four counts of disgraceful conduct and two counts of professional misconduct by the Medical Council; Green himself was found to be unfit on medical grounds to appear and so was not disciplined.

They were all led to believe that Green was simply carrying out normal diagnostic and treatment procedures.[4]

Secondly, although the project was discussed at various times by several committees, including the hospital's medical ethics committee, and although a special committee set up to investigate the complaints from some colleagues, there was a total failure by these bodies to deal effectively with the concerns raised or to seek expert opinion on the validity and safety of the research. Given this complete *lack of independent ethical scrutiny*, one of the major outcomes of the Inquiry was a complete reform of the ethics committee system in New Zealand (details can be found in Campbell 1990).

Thirdly, the Inquiry raised major questions about the scope and validity of the concept of *'clinical autonomy'*, especially when this is used to justify clinical research with a practitioner's own patients, without oversight and control by an independent ethics committee. It was obvious that Green's patients trusted him completely. They were quite unaware that he was stubbornly following a fixed belief about the insignificance of CIS, in the face of evidence that what he was doing was causing them unnecessary and, in some cases, fatal harm.

Here we see, in the clearest possible light, the severe ethical jeopardy entailed in a 'lone ranger' approach to biomedical research. Green regarded himself as a man with a mission, and seemed to genuinely believe that what he was doing was going to be to the benefit of his patients and all other patients in the same situation. The judge described him as 'a person of strong views, impatient with criticism and with total confidence in his own judgment' (Cartwright, 81). This is the nub of the problem of relying on the ethical integrity and self-determination of the individual researcher. One can imagine beneficial and largely safe research being carried out by (maybe) the majority of well-intentioned but unsupervised researchers. But the stakes are too high. We cannot leave it to the hope that all scientific researchers dealing with human subjects can be trusted to be 'good chaps'! The autonomy and welfare of all participants in biomedical research must be paramount. External review is essential.

But, while this is clearly a fundamental feature of research ethics today, is it fair to judge past research on this basis? I now turn to that question.

Judging the Past

When Philip Bonham was brought before a Tribunal of the NZ Medical Council, in light of his failure to take action over Green's research, his defence counsel argued that research ethics was relatively undeveloped in the 1960s, and so it was unfair to judge the events at National Women's Hospital on the basis of later ethical

[4] At the Inquiry Green tried to claim that he was not doing research, just trying out innovative treatment, but this was clearly unsustainable.

requirements for clinical research. I was one of the medical ethics expert witnesses asked by the Council to respond to this. For a philosopher like me, this is a fascinating question, since it raises one of the fundamental issues in moral philosophy – what aspects of ethics are universal and what are relative to time and place?

It is true that in the time of WW2, and in the two decades after it, unethical medical research was thought to be an aberration of a few doctors caught up in the murderous mission of the Nazi regime. The horrors revealed in the Nuremberg Trials were seen as totally removed from all the beneficial and trustworthy efforts of medical researchers elsewhere, whose efforts were achieving the conquest of many major diseases. There was an air of complacency about the rectitude of the research community that it would take some time to dispel. As late as 1974, the Chairman of the Ethics Committee of the British Medical Association could write this about medical collegiality: "In the relations of the practitioner to his fellows, while certain established customs and even rules are written and must be written, the principal influence to be cultivated is that of good fellowship. Most men know what is meant by 'cricket' and the spirit of the game" (British Medical Association 1974).

Of course, we now know that such reliance on everyone 'playing the game' was totally misplaced. Some doctors were involved from the war years onwards in research that, while never as extreme as that of the Nazi doctors, nevertheless showed a callous disregard for the life and wellbeing of the unfortunate research subjects.[5] Thanks to the work of H.K. Beecher, and later M.H. Pappworth, there was clearly a strong awareness from the early 1960s onwards that all was not well with mainstream medical research. Thus, documents began to circulate internationally, which were to lead in 1964 to the first version of the WMA Declaration of Helsinki.

In light of this, Green's total disregard for some of the basic provisions of the Nuremberg Code (and the fuller elaboration of these in the first version of the Declaration of Helsinki in 1964) is not excusable, given that his research started in 1966. But even if we go to the earlier era, when Mellanby was carrying out his research, we have to go below the surface of the customary rules of good relationships among medical and scientific colleagues, to the basic values on which their work was founded. In other words, we have to distinguish between etiquette and ethics. Green was not stopped for many years by his colleagues because of the rules of etiquette, still powerful at that time, which discouraged interference by concerned colleagues in the name of collegiality. It was hard even for colleagues of equal consultant rank, and virtually impossible for those more junior to him, to question the ethics of his project. Yet the underlying ethical value, whether or not it had been written in a code by then, was crystal clear, and had been so throughout the history of medicine. It is captured in the 1948 WMA Declaration of Geneva: "The health of my patient shall be my first consideration."

Mellanby was not a member of the medical profession, but it is clear from his writing that he shared this ideal and was fully committed to carrying out research to

[5] The most notorious of these was the Tuskegee Syphilis Study (Jones 1993), but this was by no means the only example of its kind.

benefit the lives of those suffering from a debilitating and unpleasant condition. Unlike Green, he did not stick stubbornly to one hypothesis in the teeth of contradictory evidence, but clearly sought to modify his research when faced with negative results. On the other hand, some doubt remains about the harm he may have inflicted in some of his later experiments, studying the natural history of the disease. He does, to an extent, admit some hesitation about the ethical acceptability of some aspects of his work: "I even found myself developing a 'conscience' regarding the sort of experiments we should, or should not, carry out" (Mellanby 1973, 51).[6] However, the main question that must be raised about Mellanby's approach to research is his persistent refusal to see any merit in independent review and oversight of his research project. Again, we have the issue of ethics versus etiquette. In the social circumstances of the time, he could no doubt easily earn the trust of those to whom he appealed for funds and other support (such as co-operation with the armed forces). He would be seen as the 'right sort of chap' who could get on with the job on his own admirably, and he had good social connections to facilitate this support. But where is the ethics in regarding your own judgment about such serious matters as degrees of risk or sharing of information with participants as needing to be exempt from independent review and advice? It is true that there were no regulations or guidelines at the time to *compel* him to seek this kind of oversight. However, the ethical value of being open to constructive criticism as a corrective to personal bias is surely a universal one, and especially strong in the culture of scientific research.

I would argue, then, that in both Mellanby's and Green's cases (but much more seriously in the case of Green) it is entirely justifiable to see them as responsible for failures in ethical conduct. It continues to be vital to stress the fundamental values in biomedical research all these years later, since the tendency to see ethical oversight of an individual researcher's work as merely intrusive and unnecessary is unfortunately still alive and well in some sections of the scientific community today. It is sadly rare to read an article by a researcher praising the work of ethics committees, and all too common to read attacks on their relevance and efficiency.

Conscientious Objectors as Research Volunteers

I come finally to the use of conscientious objectors as volunteers in biomedical research. These are the men (women were not included in his research) whom Mellanby struggled to see as 'normal'. I shall begin by talking about my father's experience in deciding to be a conscientious objector. Although this was in WWI (he saw Hitler as so evil that he was able to accept that both my elder brother and one of my sisters joined the Forces in WWII), the agonies of choice he went through would be very similar to those of many conscientious objectors in the Second World

---

[6] It is intriguing that he needs to put the word conscience into inverted commas!

War. I shall discuss different perceptions of conscientious objectors – as cowards, shirkers, less than men, versus as deeply moral, true patriots and heroes. Here I rely heavily on Lois Bibbings' excellent study, *Telling Tales about Men: Conceptions of Conscientious Objectors to Military Service During the First World War* (2009). Finally, I shall ask how all of this relates to the ethical issues of using conscientious objectors as research participants.

'A Revolution of Feeling and Mind' – A Personal Memoir

My father wrote a letter to his parents in November 1915 describing the "revolution of feeling and mind" that he had to share with them. The letter was found among his papers many years after his death, and it describes vividly the agony of choice he was faced with when he began to realise that his Christian belief compelled him to become a pacifist and not enlist in the armed forces, despite the fact that so many of his contemporaries had done so, including his own brother. The agony was made more intense by the fact that one of his cousins had already been killed in action. Dad was at the time studying for the Ministry of the Free Church of Scotland at New College, Edinburgh, and he was lodging with relatives of his dead cousin. Another very painful part of his situation was that his parents were medical missionaries, and were thousands of miles away in South India. His only means of communicating with them was by means of this eight-page letter, which would not arrive until Christmas, ironical he felt, given that such a joyous time would instead be "sad for many a heart."

It is still hard for me to read this letter, even after all these years since his death in the 1970s, as it conveys so vividly and eloquently the depth of his moral dilemma. (He was a fine preacher in his day!) He had become convinced that the Christian Gospel is unambiguous in its opposition to war. He sees as fundamental the biblical text, "Resist not evil, but overcome evil with good," and for him all that he can do is follow the example of Christ, who obeyed this precept right up to his own death. Yet, on the other hand, he feels "irresistibly drawn to protect home and country at all costs," and he knows that if his brother were killed "my first feeling would be one of red hot revenge." (Fortunately, his brother, Arthur, survived the war and worked for many years as a GP in Glasgow.) Dad's mental and emotional struggles were made still harder by the reactions of the family members with whom he was staying and later by his uncle, the father of his cousin, Tom, who was killed in the war. His attempts to explain why he had become a pacifist were met with derision, incomprehension and anger. He was told by the aunt in whose house he was lodging (he soon found he had to leave): "how it was all very well for me to stand there talking but had not Tom died for me; had I not the courage to go and fight too? – I saw she felt I was a coward." From his uncle, he received two 'very strong' letters, urging him to have nothing to do with pacifist views, but "to be a man and enlist."

Sadly, we have no record of how his parents replied to this agonised letter, and my father did not want to talk about those years, though he never regretted his decision. In the letter he says he feels sure that his parents will see his sincerity and not think that "I am trying to hide cowardice behind the cloth of Christianity, as has been hinted to me." It seems likely that his parents, with their strong Christian commitment, did understand and supported his decision to work, until the end of the war, for the Church of Scotland Huts and Canteens, which served the troops but was non-combatant.

Perceptions of the Conscientious Objector

My father's personal moral crisis illustrates one common characteristic of many conscientious objectors, a conflict between their religious belief and the strong call of other loyalties, to family and to country. However, it is important to see that conscientious objectors, in both wars, were really quite diverse in their beliefs and in their reasons for refusing to enlist. Bibbings summarises this very clearly:

> [Conscientious objectors] were a diverse group with vastly different backgrounds, views and politics. 'Conscience' could cover a whole range of motivations, including Christians, who took the commandment prohibiting killing to mean they could take no part in either the military or the war effort and men who embraced the ideals of international socialism refusing to fight their fellow workers. Others would not accept the authority of the state to direct their actions, resisting any form of compulsion, including conscription (2009, 3).

Equally diverse were the public perceptions of conscientious objectors. Probably the most common (at least in WWI), was the reaction experienced by my father – conscientious objectors were cowards, shirkers, less than men, hypocrites hiding behind the cloak of religious or moral belief. The idea of 'being a man' is especially powerful here, as Bibbings signals with the main title of her book, *Telling Tales about Men*. This association between being willing to fight and being a man has powerful sexual overtones. Virginia Woolf captured it well in a short essay written during an air raid on London in the Second World War, 'Thoughts on Peace in an Air Raid'. She writes: "We must help young men to root out for themselves the love of medals and decorations. We must create more honourable activities for those who try to conquer in themselves their fighting instinct, their subconscious Hitlerism. We must compensate the man for the loss of his gun" (2009, 5).

As Woolf suggests in this last sentence, physical aggression is easily confounded with manliness, the gun as a symbol of the erect penis. In this atmosphere the conscientious objector who refuses to fight is a cissy, more like a woman than a man, sexually inadequate. This may partly explain the derision often experienced by conscientious objectors, their manhood under a cloud, being the objects of suspicion unless they can prove themselves worthy of being seen as a 'normal' man.

Yet there are also perceptions on the other side. Conscientious objection as a different kind of valour in wartime is perhaps more recognised now than it was in the heated atmosphere of the two world wars. So far as WWII is concerned, more is now emerging about some of the horrors of war perpetrated by the Allies, something which can be acknowledged without denying the obscenity that was part and parcel of Hitler's war. Notable examples would be the block bombing of German cities towards the end of the war by the RAF, and the utter destruction of Hiroshima and Nagasaki and their inhabitants by the USAF using the newly invented atomic bomb. A notable sign of changing attitudes to conscientious objectors in more recent times has been the creation in 1994 of a memorial to conscientious objectors in Tavistock Square Gardens in London. But admiration for conscientious objectors was also evident during both wars. Bibbings quotes from this speech in the House of Commons in 1917 by an MP, Commander Wedgewood, who had served in WWI: "I think I am prouder of my country than I was before, because it has produced people who have sufficient conscientious scruples to enable them to face a long term of imprisonment rather than upset their consciences. It is something to be proud of even to produce martyrs of this sort as well as martyrs on the battlefield" (2009, 199–200). Similar sentiments were expressed in an article in *Atlantic Monthly* published in 1916: "...these men have proved their convictions genuine. They have shown themselves possessed of a moral courage at least equal to the common soldier's, and far greater than most of us educated people could show" (Bibbings 2009, 201).

In summary, then, we must avoid simplified generalisations either about the characteristics of those who have chosen to be conscientious objectors or about the public perception of their actions. The picture is a complex one, but one feature is clear: Virtually all conscientious objectors have had to show considerable courage in following the dictates of their conscience and thereby being willing to risk both threats to their personal freedom and – most commonly – threats to their self-esteem through social derision and ostracism. In times of war no-one choses to be a conscientious objector lightly or as an easy way out of danger.

The Use of Conscientious Objectors as Volunteers in Research

I come finally to the issues surrounding the use of conscientious objectors in research. The key issue here revolves around the term 'volunteer'. As we saw earlier, volunteering must be seen in the context of the social circumstances in which the invitation to volunteer is given. Mellanby recruited most of his volunteers from the Pacifist Service Units. These were set up by the Peace Pledge Union to provide employment for conscientious objectors who had been granted conscientious objector status by a Tribunal, but on condition that they undertook designated employment instead of doing military service.[7]

One important point to note here is that this group of volunteers had made a case for their conscientious objection that passed the fairly rigorous scrutiny of a Tribunal (18,000 of the 61,000 persons seeking conscientious objector status in WWII were

[7] Similar provision for this group of conscientious objectors was made by the Friends' Ambulance Units, but these were for more specific tasks related to hospitals.

refused it, and were required to serve in the armed forces or go to prison. Of these, 5000 men and 500 women were imprisoned.) Moreover, they had not received *unconditional* exemption, which was granted only to 3000 conscientious objectors by Tribunals, but were obliged to undertake an approved occupation. These circumstances make them an unusual class of volunteers in several ways, since their choices were limited. The most obvious constraint on their choice was that there was fairly restricted range of occupations from them to chose from. It is possible that the conditions offered by Mellanby, which included board and lodging plus a small allowance, would have been an inducement to those who did volunteer. We are not told anything about their personal circumstances in the book, but presumably these were men without families to live with and support, thus making the residential offer an attractive and convenient one. We are told that some of them stayed on for several years and were used in a variety of projects, which does suggest an advantage to them compared with other approved occupations.

However, a much more important constraint on their choice was the strength of these men's consciences. I mentioned above the way in which being a conscientious objector in wartime can pose a significant threat to one's self-esteem, given accusations of cowardice and selfish regard for one's own safety. For these volunteers, being able to do work which would be of real service and could help overcome a widespread and highly unpleasant condition affecting both children and adults, must have seemed a very clear and practical way of showing that they still had a genuine concern for others and for the welfare of their country. So, we have here an obvious way in which this was a group much more likely to agree to be part of a medical research project than the population at large. It may be too much to describe them as a vulnerable group, but certainly they can be seen as a group much more susceptible to moral persuasion.

What can be concluded from this in terms of recruiting conscientious objectors into biomedical research projects? It can be argued that there is no ethical objection to approaching them with a request to volunteer, since they are clearly not compelled in any way to agree to this as their choice of designated occupation. The concern is more with *continued* participation in projects, like Mellanby's, where the risks are increased. Refusal for this group becomes a *second* refusal to subject themselves to physical dangers, but now danger for an apparently worthy cause. It is really no surprise that Mellanby found his subjects to be willing, or even keen, to subject themselves to all sorts of unpleasant experiments, including ones that had a risk to their health. How could they say no, without been seen – by themselves and maybe also by others – as cowards after all? For this reason, great caution should be exercised in recruiting conscientious objectors into research projects, and, as I have said several times already, effective independent ethical research governance is absolutely essential. This is the main lesson that can be drawn from a study of Mellanby's *Human Guinea Pigs*.

References

Beecher, H.K. 1970. *Research and the individual*. Boston: Little, Brown.

Bibbings, L.S. 2009. *Telling tales about men*. Manchester: Manchester University Press.

British Medical Association. 1974. *Medical ethics*. London: British Medical Association House.

Campbell, A.V. 1990. Ethics after Cartwright. *New Zealand Medical Journal* 104 (905): 36–37.

Cartwright, S.R. 1988. *The report of the committee of inquiry into allegations concerning the treatment of cervical cancer at National Women's Hospital and into other related matters*. Auckland: Government Printing Office.

Jones, J.H. 1993. *Bad blood, the Tuskegee Syphilis experiment*. New York: Maxwell Mcmillan.

Manning, J. 2009. *The Cartwright papers*. Wellington: Bridget Williams.

Mellanby, K. 1973. *Human guinea pigs*. London: Merlin Press.

Pappworth, M.H. 1967. *Human guinea pigs: Experimentation on man*. London: Routledge and Kegan Paul.

Ten Have, H. 2016. *Vulnerability: Challenging bioethics*. London: Routledge.

Woolf, V. 1952. *Thoughts on peace in an air raid*. London: Penguin Books.

Chapter 21
Compensatory Service for Conscientious Objection

Companion Essay for Mellanby Reprint

Joseph S. Brown and Toby Schonfeld

In *Human Guinea Pigs* (1945), Kenneth Mellanby argues that not only is research participation appropriate compensatory service for those who claim conscientious objection to combat during war time, but that research participation could/should be a national effort. Towards the end of the text, he even recommends a government bureau devoted to this pursuit.

Upon reading Mellanby's arguments on conscientious objection, alternative service, and research participation, we were struck both by how similar these issues were to contemporary versions of conscientious objection, and also by how differently these issues are being resolved today. From this observation, we believe it might be fruitful to explore the logic of conscientious objection as it was practiced in Mellanby's time and in the context of combat. We will also attempt to apply some of the concepts underlying those practices to modern instances of conscientious objection on behalf of heath care providers and other members of society.

The connection between combat objection and participation in clinical trials might seem odd at first, but in fact conscientious objection and the provision of health care have often been linked. That is, conscientious objection to combat is often linked to the provision of medical service as a combat medic or, in the case of Mellanby, research participation. In that spirit, Mellanby's suggestion about research participation as compensation for conscientious objection to combat provides a

J. S. Brown
Department of Psychology, University of Nebraska-Omaha, Omaha, NE, USA
e-mail: josephbrown@unomaha.edu

T. Schonfeld (✉)
National Center for Ethics in Health Care, US Department of Veterans Affairs, Washington, DC, USA
e-mail: tschonfeld@primereviewboard.com

L. M. Rasmussen (ed.), *Human Guinea Pigs, by Kenneth Mellanby: A Reprint with Commentaries*, Philosophy and Medicine 134,
https://doi.org/10.1007/978-3-030-37697-0_23

mechanism for us to explore whether compensatory service is a necessary (or perhaps desirable) component of conscientious objection. Then, we will explore the necessary components of required compensatory service for conscientious objection to combat. We will expand to a brief overview of conscientious objection to accepted practices in health care. Although it is a bit outside a narrow interpretation of this topic, we will also consider a few instances of how these solutions might play out in more diverse cases of conscientious objection.

Setting the Stage: Definitions and History

Mellanby's thoughts about research participation as a way for conscientious objectors to still "serve humanity" came about during the prelude to World War II (Mellanby 1973, 22). While not a pacifist himself, Mellanby was sympathetic to those who were unable to participate in war and yet wanted to contribute to society in some meaningful – not menial – way. To him, "volunteering" for research participation was a perfect solution, so long as the research "… was not solely directed to improving military efficiency" (p. 23). Rather, research with a broader goal in mind would likely be viewed favorably by objectors so long as conscription continued.

To that end, Mellanby proposed "… a special Institute for the study of 'Human Biology.'" After some very specific ideas about where in the United Kingdom such an institute could be physically housed, Mellanby goes on to describe the nature of the work and the conditions that would need to obtain were his proposal to move forward (1973, 111–2):

> Such an Institute would require a small but carefully chosen permanent staff of medical and scientific workers and a permanent nucleus of volunteers. There will always be a small number of men who, with reasonable prospects for the future, will be willing to make a career of this type of work and who will be invaluable in teaching new subjects the necessary principles of human experiments. As a whole, the volunteers should be recruited on a short-term basis, for one particular experiment which may last a month or up to six months….
>
> Assuming that such an Institute came into being, some scientific work would be in progress all the time under the direction of the permanent staff. But it would also serve as a centre where scientists from all over the country, or the world, could come for a period to work on their own problems. In the same way in which marine botanists or zoologists who are marooned at inland universities go to the laboratories at Plymouth or Millport during vacations to work on their own problems, so human physiologists and medical workers could come to our Institute….
>
> I am sure that the project as sketched above would be feasible, provided that the funds were available. The two questions which would need to be decided relate to finance and to control. An Institute for the study of human biology would not be an inexpensive proposition, though the money involved may seem very little when compared with the cost of the war. The question of control is more difficult; there seems no doubt that it would have to be in the hands of a committee of scientists, but their choice would present many problems, for they would not only require a considerable knowledge of the scientific problems involved, but also the ability to understand and manage the volunteers. In the long run, however, I think the success or otherwise of the venture would depend on the choice of the small

> nucleus of permanent volunteers, whose work it would be to see that the correct sort of tradition was maintained, so that newcomers felt they were entering an organisation where they could really be of use.

In order to fully assess his proposal, we first need to consider the fundamental role that conscientious objection plays: objecting to war-related duties is a necessary foundation for all that follows for Mellanby. Many accept at face value that no one should be required to act against one's conscience. This commitment is most often grounded in the respect for an individual's moral integrity, but also can be based on additional concepts: respect for autonomy, respect for moral/cultural diversity, and the notion of epistemic humility – the recognition that we may not be right about what we consider ethically justified, and so respecting others' approaches to the moral life acknowledges that their actions may be closer to the "truth" than are ours (Wicclair 2011, 25–29). This has resulted, in the words of John Whiteclay Chambers II, in a culture that emphasizes "...the need for a proper balance between the rights and sincerely held beliefs of the individual and the interests of the community" (Chambers 1993, 23).

Conscientious objection can take many forms. Genuine conscientious objection consists in more than simply saying or thinking "I don't like that idea" and acting accordingly. Conscientious objection is not simply a preference claim; more is required to justify that a claim is truly an exercise of conscientious objection, and there may be corresponding obligations that accompany the exercise of conscientious objection (LaFollette and LaFollette 2007). Conscientious objection is also not merely an exercise of professional integrity, where one refuses to perform some action because it is beyond the scope of her expertise. An act of conscientious objection implies three things on behalf of the agent (Wicclair 2011, 5):

1. The agent has "core" values that can have moral or religious grounding;
2. Those values contribute to the essential nature of the agent; that is, they define who he/she is on a fundamental level;
3. Performing the action violates the individual's "core" values, and therefore proceeding would be "self-betrayal" and may lead to a "loss of self-respect."

In general, conscientious objection involves excusing actors from certain *prima facie* obligations. In the most commonly discussed instances of conscientious objection – conscientious objection to participating in war – dissenters are not permitted to "opt out" simply by stating an opposition to action. Instead, conscientious objection is what Hugh and Eva LaFollette (2007) call a "qualified right": a right that obtains when certain conditions have been met. In the case of the draftee, for example, those conditions include proving to an independent body that: (1) the draftee has "core" ethical values that are (2) central to the individual's self-concept, and that (3) participating in certain activities – in this case, combat – are fundamentally inconsistent with those essential values. Furthermore, those who argue successfully for conscientious objection to war are obligated to participate in some amount of other "civil service" in lieu of military service; conscientious objection to combat is not consequence-free.

Conscientious Objection in Health Care

Even if there are good reasons to respect some conscience claims, there remains a gap between what constitutes respecting these claims in the health arena and what constitutes respecting them elsewhere. In an effort to demonstrate the lack of parallelism between conscientious objection to war and conscientious objection in health care, Stahl and Emanuel argue that acts of conscientious objection in health care fail to meet any of the constitutive elements of conscientious objection to military service (2017). In their view, the five defining features of conscientious objection to military service are that: (1) it presupposes – and opposes – state-mandated conscription; (2) it "opposes an unchosen combatant role;" (3) "it requires all-or-nothing…objection;" (4) it requires judgment of the objection by a third party; and (5) acceptance of the objection entails either compensatory service or imprisonment.

When observed from this lens, the problems with the analogy are apparent. For one thing, there is no such thing as "mandatory" provision of health care; indeed, precisely because individuals voluntarily choose to enter clinical medicine, some have argued that this choice virtually removes the foundation upon which conscientious objection is tenable: if you do not want to provide some good or service, then choose another profession (Stahl and Emanuel 2017). Others have argued that conscientious objection needs to remain a viable option in order to preserve providers' integrity when, especially, they are faced with new interventions that they could not have foreseen when they entered the profession. (Consider, for example, the emergency department provider who had no notion that a "morning after pill" might come within the scope of her practice). Similarly, it might be argued that excusing a provider who objects to a certain procedure on moral grounds protects patients from inadequate care (Wicclair 2017). Regardless, it looks like the first two features of conscientious objection – opposition to mandatory conscription and opposition to mandatory combat – are missing from health care exercises of conscientious objection.

Yet the literature is replete with examples of health care professionals claiming conscientious objection. Examples of such claims recur in response to the provision of reproductive care (abortion, emergency contraception) (Harris et al. 2018; Morrell and Chavkin 2015), to vaccination practices (Clarke et al. 2017), to the provision or reduction of opioids for pain management (Magelssen 2017), to participation in medical aid-in-dying processes (D'Angelo et al. 2019), and to providing blood products (Hudgins and Carter 2019), just to name a few.

Even though we cannot fully explore the connections between claims of conscientious objection in health care and those of military service here, we are intrigued by a relatively recent approach to applying two symmetrical criteria for conscientious objection in health care: that of third-party judgment of the conscientious objection claim and the application of compensatory action (criteria 4 and 5 as described by Stahl and Emanuel). Scholars have begun exploring the use of a "tribunal" to vet a provider's claim of conscience (Hughes 2017; Clarke 2017). The idea is that a group of independent assessors would examine the objector's claims and rationale and determine (a) whether or not the claim should be supported in this instance, and if so (b) what compensatory actions are appropriate.

This explicit consideration of compensation for conscientious objection in health care is new; previous approaches have simply argued about the justification (or lack thereof) for respecting conscientious objection and what legal and/or public policy implications may flow from these positions. In this way, conscientious objection in health care has been consequence-free to providers who claim it, but – importantly – not consequence-free to the patients who are harmed because of institutional respect for this right. Attending to the notion of third-party assessment and compensation restores a sort of balance to the health care equation, where those who refuse to perform some action would be required to, instead, perform something else on behalf of (if not this patient, then) others. This is the sense in which we see analogies to Mellanby's suggestions of research participation as compensation for conscientious objection to combat, and we wonder what that would look like if applied to health care broadly. There is historical precedent for the evolution of the concept of conscientious objection, as well as its appropriate alternatives (Chambers 1993). It is into this tradition that we submit what follows.

Alternative Service

The introduction of the analogy of tribunals[1] – a constitutive feature of the military model – provides a fruitful ground for our consideration of research participation as alternative service. In the military model, the tribunal's job is to ascertain if the objector has moral commitments that oppose (say) *all* killing, or just his/her/their active engagement in killing (Clarke 2017). Once convinced of their sincerity and authenticity, the tribunal then decides whether noncombatant (but still military-related) roles are appropriate compensatory service, or whether other forms of national service will suffice (Weiss 2012). Thus, even if you have objection to the use of a tribunal in determining the validity of claims of conscience, there would need to be some mechanism to assess compensatory service. Some sort of third-party judgement is required.

Both in the context of military service and healthcare it is important to note that the alternative service required from those who have an objection of conscience is not appropriately viewed as punishment. Mechanisms to punish those who refuse society's wishes abound, and in cases of refusal of military service, those mechanisms (often execution) are applied to refusals to comply. However, the rationale for

[1] We are both disturbed and intrigued by calling these boards "tribunals." The allusion to the military model is obvious, and in that sense we worry about the connections between concepts of order, duty, and hierarchy that are necessary features of the military. On the other hand, to the extent that the tribunals would apply across the board, regardless of position or purpose, we are intrigued by the uniformity suggested by the term. Since we are simply exploring the concept here as it relates to research, we will adopt the term and its multi-faceted implications for the purpose of this essay.

accommodation of a conscientious objection is not punishment but restitution for withholding a mandatory duty. We do not punish the refuser, for their action is (at least by their lights) moral – only absolutism (no service, no compensatory action) remains grounds for punishment (Chambers 1993, 24). Further, as a society, we have accepted the notion of individual morality overriding public duty, at least in some circumstances. Still, those who refuse military service have withheld from society that which they owe. In order to balance the scale, some equivalent act is required.

We note that this notion of restitution is at least compatible with a number of deeper philosophical positions that might be relevant. Whether one prefers consequentialist style arguments where harm must be mitigated or offset, more deontological notions of justice, or more contractual conceptualizations of justice, restitution might be seen as a valid remedy for the failure to perform a mandatory action.

It is also the case that the requirement for restitution serves a practical objective: it reduces the incentive to claim conscientious objection merely to avoid an act you would prefer not to do. Anyone of sufficient wit can manufacture an ethical objection to virtually any act and, should conscientious objection become a consequence-free option, we would be buried in such claims. This would be especially true for very aversive options like war but might apply to a number of activities seen as unpleasant.

Of course, having established a desire for restitution, the question of equivalence arises. It seems there are several dimensions of equivalence that must be considered: equivalence of risk, equivalence of benefit, and the appropriate party to which restitution should be made.

Research as an Alternative 1: Balancing Risks

Given this framework, we will examine participation in research as an alternative to military service and to the conscientious objection of a health care provider. Such a discussion should explicate our position and explore its implication.

To embark on this discussion, we need first to consider the basis of the equivalence of the required societal act and the alternative provided in the face of conscientious objection. One reasonable dimension to consider would be the level of risk. In conscientious objection for military service, we see a case where risk seems particularly relevant. The act of becoming a soldier presents a host of risks/costs. One must leave one's home and job, engage in dangerous training and then (often) go even farther from home to engage in even greater risks. Loved ones left at home also experience risk: financial, emotional, etc. One can easily see the incentive to avoid such actions. Thus, serving as a combat medic is often offered as an alternative to direct action in combat. As the medic is in the field with the war fighters, the one claiming conscientious objection shares the risks and costs of combat. This includes both the direct risk (people trying to kill you) and the indirect risks/costs associated with training and deployment.

Research participation comes with its own set of risks. They are study-dependent, but range from a mere loss of time to significant morbidity and even mortality. Some research presents emotional or psychological risks and some presents the possibility of loss of confidentiality or privacy. Some research also presents risks to those not involved in the research: genetic research often reveals (sometimes damaging) information about family members who did not agree to participate in the study. And also like the vagaries of war, research brings with it unknown risks: if we knew all that we could expect, there would be no need to do the research in the first place. All of this is to say that it would likely be possible to compare risks of combat to risks of research as one means to balance the scale for objectors.

Although there are practical reasons to consider risk, we would suggest that risk is neither necessary nor sufficient for compensatory service. Were it sufficient, one could replace military service with a sufficient time playing in traffic or playing Russian roulette: risk is risk, after all. Further, it seems that given the justification we have provided – that research participation is compensation for a withheld service – risk comparisons should not be necessary either. So long as society receives equivalent value, risk should be irrelevant. It is only when you consider conscientious objection as a punishment that equivalent hardship might be relevant.

Further, we would argue that using risk as a component of our analysis brings some serious moral dangers into play. If the conscientious objection is to a particularly dangerous activity like going to war, then to balance the risk and hardship, one would either have to engage in extremely risky or extremely unpleasant activities whether that risk was beneficial or not. That is, the risk would be a required component of the replacement act. In the case of research participation as a possible alternative service, this would breach the fundamental agreement that we have with subjects: that we will not subject them to needless harm. This is particularly true given the experiences Mellanby cites, where the research volunteers seemed "…to be tougher and more willing to suffer danger than most other members of the population" (p. 127).

This is not to say that risk is entirely irrelevant. We would argue that risk is relevant only indirectly. That is, one component of the alternative action is it must not be so attractive as to induce great numbers to court this alternative service. That is, even if spending a week playing golf brought about some tremendous benefit, it would likely induce so many people to apply for this alternative, that it might be impossible to recruit a fighting force.[2] This practical concern might be partially alleviated by equating the time served in the two endeavors. Although a year of providing health services to underserved individuals might be less dangerous or

[2] We will leave aside the moral implications of a war that is insufficiently supported by the populace to allow for a volunteer fighting force.

unpleasant than a year in theater, equivalent time commitment does go some distance in reducing the incentive to claim conscientious objection status.[3]

Research as an Alternative 2: Balancing Benefits

We would now like to consider another dimension upon which service might be deemed equivalent: the accrued benefit. Recruiting our example from above, service as a combat medic is often used as an alternative to direct involvement in combat. Above, we noted that this satisfied an equivalent risk metric, but there is another feature of serving as a combat medic that is worth noting: it replaces the withheld service (combat) with a like benefit to society (the advancement of whatever goals that society sees as the objective of the war). Additionally, it provides benefit to the class of people who have been harmed by the refusal. This class includes society (that loses the fighter) and those who have assumed the rejected duties. Thus, the class "soldier" has sustained higher risk of combat because of the action of the person claiming conscientious objection, but also has received an amelioration of this risk from the objector's alternative actions.

Similarly, research participation as an alternative to combat might be restricted to research objectives that would benefit the war effort or improve the lives of soldiers as a class. This was at least partly the case for those instances described by Mellanby. In the broader context, this restitution to the damaged group provides a particular benefit in that it allows the separation of the required act that is perceived as immoral by the objector and the persons for whom the act is performed. History is full of examples of research pursued to address the needs of soldiers that involved non-combatants who were significantly harmed by this work (Baker 2020, this volume). That harm certainly cancels out any compensatory benefit on the scale. Yet if the goals of that work were truly to enhance the lives[4] of members of the military, then it seems as though objectors could have engaged in this research as appropriate compensation for service

Applying this analysis to refusal to provide care, we might consider a therapist's refusal to provide counseling to a client who presents as transgendered. By considering both society's need to have productive healthy members, and the transgendered community's need for psychological services (which they share with all other segments of society), we might begin to see an outline of an appropriate alternative service. In this context, society might be served by a broad range of mental health services, like providing services at a free clinic. Since society's need is for

[3]Please note that issues of the attractiveness of the alternative service addresses the *nature* of the compensatory service, not its moral necessity. Even if you find a service (e.g., providing therapy or baking a cake) quite pleasant, you still owe compensation if you refuse to perform this service when required by society.

[4]Actually, it was likely to return soldiers to fighting form, but a full discussion of this is beyond the scope of this essay.

psychological services, it really doesn't matter to whom they are provided (so long as the service was not provided by other means, like paying a clinician). However, the other injured parties include the particular patient and the transgendered community in general. The alternative service should serve *this particular* community, rather than society's overall needs. It would therefore be insufficient to consider (general) free clinic service as appropriate compensation. Rather, something like professional contribution to efforts to reduce violence or bullying against transgendered people might be appropriate. That way, the therapist would not be engaging in the service that the therapist views as problematic but could still contribute to the welfare of the community.

Challenges to Our Proposal

Considering compensatory service as a mandatory element offers a number of advantages, but there are also practical issues at stake. Should one adopt our proposal, then an inordinate burden might be forced on objectors as a form of retaliation. For example, having refused to provide a cake for a local police fundraiser (perhaps because of concerns about racial disparities in policing), every police officer from miles around might come to demand that service from *this particular* baker, thus forcing upon them a level of compensatory service that might very well be incompatible with running a business or pursuing one's own life. However, there are ways to practically limit such concerns. For instance, in the case of a baker, one might issue a refusal for all of a certain type of cake. One's level of restitution could then be based on the proportion of the market you serve. That is, if there are 1000 cakes of the offending type ordered in your city each year, and you are 5% of the cake market, your restitution could be the equivalent of 50 cakes.

Additionally, in some cases, the issue of to whom to provide restitution becomes a bit more difficult. Given the nature of the objection, our baker might equally object to a number of alternatives that benefit the correct group (the police officers). However, with a bit of consideration, there are reasonable alternatives. For instance, the baker could supply the cakes to a fundraiser for the dependents of slain officers. Note that this approach requires that the person refusing their duty must offer specific reasons and accept reasonable alternatives to that service that accommodates those reasons. Thus, it prevents animus toward a given group (in this case the police) hiding behind a much more publicly acceptable reason (disparate policing). The alternative we suggest allows relief from the more focused objection but does not allow relief from the larger class-based consideration (in that the restitution must be to the community that was denied the service).

We offer this example not to solve all the possible practical problems, but to illustrate that they are solvable. Those solutions are not without costs, but they serve the goal of allowing the respect of a conscientious refusal while mitigating its harms to the individuals who are refused a public service. Further, it reduces the incentive to refuse public service on a whim.

We also note that this discussion does not resolve the issue of what behaviors are mandatory. Whether one is required to offer public accommodation or housing to anyone who asked, whether opening a business requires you to serve all customers, or whether it is morally appropriate to draft citizens to combat are all separate issues from how to accommodate conscientious objection to these putatively mandatory duties.

Returning to Mellanby

In the case of research participation as an alternative to mandatory combat duties, we see a reasonable resolution to both the costs of forced participation in an act that is seen as immoral, but also the cost of the refusal to act.

We find Mellanby's example so compelling because it is such an obviously applied version of the role of conscientious objection. A society at war must honor its moral foundation, but it cannot do so at the expense of the practical considerations of war. The cost of conscientious objection was obvious and meaningful in Mellaby's case. In contemporary discussions, it is easy to overlook the consequences of conscientious objection because the consequences are diffused across many individuals. Further, the nature of conscientious objection in contemporary society means that the individuals who are harmed are of lower social status[5] than the person refusing service. It may be for this reason that the notion of compensatory service has, until recently, received little consideration.

Disclaimer The views expressed by this author are the author's own and do not represent the views of the US Department of Veterans Affairs or the US Government.

References

Baker, Robert. 2020. *Human guinea-pigs*: Mellanby, Pappworth and Club Regulation, this volume.

Chambers II, John Whiteclay. 1993. Conscientious objectors and the American state from colonial times to the present. In *The new conscientious objection: From sacred to secular resistance*, ed. Charles C. Moskos and John Whiteclay Chambers II, 23–46. New York: Oxford University Press.

Clarke, Steve. 2017. Conscientious objection in healthcare, referral and the military analogy. *Journal of Medical Ethics* 43: 218–221.

Clarke, Steve, Alberto Giubilini, and Mary Jane Walker. 2017. Conscientious objection to vaccination. *Bioethics* 31 (3): 155–161.

[5] We do not believe it is entirely coincidence that most contemporary cases of conscientious objection generally entail a high-status individual (a physician, a therapist, or a business owner) denying service to a lower status individual. Further exploration of this point is beyond the scope of this essay, however.

D'Angelo, Abby, Kelly Ormond, David Magnus, and Holly K. Tabor. 2019. Assessing genetic counselors' experiences with physician aid-in-dying and practice implications. *Journal of Genetic Counseling* 28: 164–173.

Harris, Laura Florence, Jodi Halpern, Ndola Prata, Wendy Chavkin, and Caitlin Gerdts. 2018. Conscientious objection to abortion provision: Why context matters. *Global Public Health* 13 (5): 556–566.

Hudgins, Kerstin, and Esther Carter. 2019. Blood conservation: Exploring alternatives to transfusions. *Critical Care Nursing Quarterly* 42 (2): 187–191.

Hughes, Jonathan A. 2017. Conscientious objection in healthcare: Why tribunals may be the answer. *Journal of Medical Ethics* 43: 213–217.

LaFollette, Eva, and Hugh LaFollette. 2007. Private conscience, public acts. *Journal of Medical Ethics* 33: 249–254.

Magelssen, Morton. 2017. Professional and conscience-based refusals: The case of the psychiatrist's harmful prescription. *Journal of Medical Ethics* 43: 841–844.

Mellanby, Kenneth. 1973. *Human guinea pigs*. 2nd ed. London: Merlin Press.

Morrell, Kathleen, and Wendy Chavkin. 2015. Conscientious objection to abortion and reproductive healthcare: A review of recent literature and implications for adolescents. *Current Opinion in Obstetrics and Gynecology* 27 (5): 333–338.

Stahl, Ronit, and Ezekiel Emanuel. 2017. Physicians, not conscripts – Conscientious objection in health care. *New England Journal of Medicine* 367 (14): 1380–1385.

Weiss, Erica. 2012. Principle or pathology? Adjudicating the right to conscience in the Israeli military. *American Anthropologist* 114 (1): 81–94.

Wicclair, Mark R. 2011. *Conscientious objection in health care*. New York: Cambridge University Press.

———. 2017. Conscientious objection in healthcare and moral integrity. *Cambridge Quarterly of Healthcare Ethics* 26: 7–17.

Chapter 22
Of Mites and Men: What WWII Scabies Experiments Can Teach Us About New, Unregulated Human Subject Research

Lisa M. Rasmussen

Introduction

Human subject research in the United States has historically been governed by federal regulations attached to grant funding support. Through this regulatory reach, it has been possible to enforce standards of ethical treatment, built on the foundations of the Nuremberg Code, Declaration of Helsinki, and the *Belmont Report*, on the vast majority of research with human subjects in the United States. However, new research methods, such as citizen science, DIY biology, biohacking, and corporate research all pose challenges to the conventional approach, because they can be left ungoverned by these regulations. This requires us to think anew about how to ensure such research is conducted ethically. How can we think about ethical research in the absence of regulatorily required or universally shared norms?

In order to answer this question, it may help to consider how research was conducted before the articulation of the Nuremberg Code, and how researchers who worked with human subjects understood and implemented their obligations to those subjects. Such studies have, for example, revealed the ethical principles of Nazi socialist medical research (Cooper [Ramm] 2019), and of the Scottish physician John Gregory (McCullough 1998). Here, I consider the work of Kenneth Mellanby, whose work on scabies transmission with conscientious objectors during WWII stands astride the old and new approaches to ensuring ethical research with human subjects, especially in light of his work as a correspondent to the Nuremberg Doctors Trial for the *British Medical Journal*. His work was thoughtful, conscientious, and caring, yet even his own description of the work reveals problematic oversights and prejudices common to any human being. Mellanby's celebration of the freedom of

L. M. Rasmussen (✉)
Department of Philosophy, University of North Carolina at Charlotte, Charlotte, NC, USA
e-mail: lrasmuss@uncc.edu

L. M. Rasmussen (ed.), *Human Guinea Pigs, by Kenneth Mellanby: A Reprint with Commentaries*, Philosophy and Medicine 134,
https://doi.org/10.1007/978-3-030-37697-0_24

the scientist to conduct "untrammeled" research – research not subject to committee direction or oversight – teaches us lessons that will help us to identify potential ethical issues in new forms of independent research.

The Human Guinea Pigs

Kenneth Mellanby gained prominence due to his scabies research during World War II, an account of which was first published in 1945 (and again in 1973) as *Human Guinea Pigs*.[1] He was born in 1908 in Renfrewshire, Scotland (Mellanby n.d.), and after studying Natural Sciences at Kings' College, Cambridge, earned a doctoral degree at the London School of Hygiene and Tropical Medicine.[2] He was subsequently awarded the Royal Society's Sorby Research Fellowship, and initially worked on "lice, fleas, bedbugs and other insects of medical importance," establishing that louse infestation was much more widespread than had previously been thought (Mellanby 1988). This work also made him aware of physicians' growing concerns about scabies.

There were several timely reasons to focus on scabies in 1941. The infestation was thought at that time to be transmitted via contact with fomites such as bedding, and was thought to be controlled via "stoving" (heating) clothes and bedding to kill the insects. German air raids were expected, which would lead to many people gathering tightly in shelters, and create more opportunities for transmission and fewer opportunities for preventive measures like stoving. In addition, in World War I, levels of infestation had kept soldiers away from duty in high enough numbers to impact troop readiness (Farmer 2014), so the military was searching for an effective means of addressing the condition in WWII. Stoving of clothes and bedding was time-consuming, costly, and not demonstrated to be effective. Thus, the situation was ripe for the attention of an entomologist who, because of his scientific expertise, had been forbidden from serving as a combatant due to the Schedule of Reserved Occupations (Mellanby 1973, 20).[3] By working on the problem of scabies, Mellanby could put his talents to good use on behalf of the country and the war effort, even if he himself could not fight.

In 1940, Mellanby proposed a study on scabies to the Ministry of Health, outlining laboratory experiments to study the life cycle of the pests away from hosts, including life history and means of destruction, and 'field' experiments testing the transmission of scabies on human beings. He also proposed using conscientious

[1] For a history of how "guinea pigs" became the term used to describe human subjects of research, see Engber 2012).

[2] Biographical information compiled from Perring 1994; Editor 1947.

[3] "The Authorities, no doubt wisely, wished to avoid the mistakes of the last war, when so many men, whose professional or scientific qualifications would later have been valuable, joined the Army in 1914, so that the country was deprived of their special abilities. In order to make the best use of all technically-qualified men and women in any emergency, the 'Central Register' was set up some time before the outbreak of war" (Mellanby 1973, 20).

objectors as test subjects.[4] At the end of 1940, he was awarded £5000 to conduct the research, secured a house for the research at 18 Oakholme Road in Broomhill (a suburb of Sheffield), and outfitted it in preparation for the studies.

On January 25, 1941, the first two volunteers arrived at the Sorby Research Institute. One month later the Institute was full of male conscientious objectors, including 12 live-ins and two who lived nearby. They were excused from combat by a Tribunal and allowed to serve in this capacity instead, and they were eager to "do their part" for the war effort.[5] Pay was lower than a combatant would have earned, but housing, food, and other subsistence was provided, and volunteers helped to keep the house and conduct some of the research.

The first sets of experiments were designed to determine exactly how scabies was transmitted. These experiments involved sleeping naked in bedding just vacated by infested patients, or wearing the just-discarded undergarments of infested patients "night and day for a period of at least a week," but they were unable to successfully transmit scabies for quite some time. Eventually, they did obtain two successful cases of infestation due to the wearing of underwear "when still warm from the body" of the previously infested patients. However, the infestation took over a month to manifest symptomatically, which was the first evidence they had that infestation did not occur in mere days. Having established some of the conditions for transmission, they infested future volunteers by harvesting *Sarcoptes* mites from infested patients and introducing them directly onto the volunteers in the desired location, which enabled them to observe the natural progression of the condition.

Mellanby identified multiple results of the experiments, in addition to the transmission pathway and timing of infestation. He observed that the body does mount some defense against infestation, causing it to be more difficult to infest people a second time, and hypothesized that this might explain the historic periodicity of scabies outbreaks. The research team established that scabies was much more widespread among the population than had previously been thought; that the war-time conditions had almost nothing to do with the infestation rates; and that the civilian population actually suffered infestation at a higher rate than the military population. Critically for the war effort, they also established that "stoving" of bedding and clothing was unnecessary for the treatment of scabies in the population. This led to a revision of the Army's "Memorandum on Scabies" to eliminate the need for stoving when patients had been treated properly, and significant savings to the Army as a result (Mellanby 1973, 117–118). Finally, their comparisons of various treatment regimens resulted in a straightforward, painless, very effective, and fairly brief standard approach.

---

[4] "Preliminary discussions had suggested that there would be a good response to an appeal for volunteers to serve as subjects for medical research, provided that it could be shown that this would be likely to be of value in alleviating suffering, and...that it was not solely directed to improving military efficiency" (Mellanby 1973, 33).

[5] Mellanby's attitude towards conscientious objectors can be gleaned from his comment that "[t] hey seemed to be more or less normal people" (48). One of the conscientious objectors, Allen Jackson, later became a "lifelong leading member of the Peace Pledge Union" (Melicharova 2006).

Mellanby's work is an important foundation for current approaches to the treatment of scabies, and was reviewed as recently as 2004 in the category of "Classics in Dermatology" (Currier et al. 2012; Jackson 2004). Individual clinicians have also noted the importance of Mellanby's work in their own clinical practice: One reports having been told in the 1960s, when he requested help in identifying scabies at the bedside, "Go away and read some Mellanby" (Savin 2002, p. 86). Another commented in 1974, shortly after the publication of the second edition of *Human Guinea Pigs*, "Of the monographs studied during my training, the briefest, best written, and most exciting was Sir Kenneth Mellanby's Scabies [*sic.*]" (Maibach 1974). But however beneficial the results of the experiments, one must always be cautious of using benefits to justify costs. The costs of Mellanby's experiments to human subjects are difficult to identify using his account alone (although in some places it is striking how baldly he describes the suffering of the volunteers). But we can contrast our current regime of human subject protections with Mellanby's to identify what would count as potential costs today.

It would be hubris to presume that we happen to live in the time where we've gotten the moral calculus right. However, what our 70+ years of considering human subject research carefully *have* given us that Mellanby lacked are publicly considered principles and values, undergirded by arguments and reasons, guiding the ethical conduct of research. This at least makes it possible for researchers, their subjects, and society to consider together the kinds of research we might consider ethical. The determination of what constitutes ethical research with human beings is no longer remanded to the conscience of the "untrammeled" researcher, or to the "more reliable safeguard provided by the presence of an intelligent, informed, conscientious, compassionate, responsible investigator" Henry Beecher celebrated in his famous article on human subject research (Beecher 1966, 1360). Contemporary ethical standards require oversight of human subject research by an independent body. However, new research methods have the potential to reinstate the "untrammeled" researcher's conscience as the sole determinant of ethical standards of human subject research. Mellanby's projects also did not receive formal external oversight, so it is instructive to consider the ethical issues evident in his work and what that might imply for new forms of unregulated research.

Lessons from *Human Guinea Pigs*

Since 1945, when Mellanby's first edition was published, research with human subjects has spawned entirely new career and business opportunities: IRB administration, for-profit IRBs, Contract Research Organizations, and even "guinea-pigging," a colloquial name given to those who regularly participate in paid research, usually for pharmaceutical research.[6] Despite the fact that Mellanby himself endorsed the idea of a government-funded permanent institute for research on human beings (1973, 155–158), given his avid support for "untrammeled" research, he likely would have been appalled by the regulatory infrastructure that gives rise to these careers. Even in the second edition of his book, written almost three decades after the first, and after he had had a front-row seat at the Nuremberg Doctors' Trial, he cited as one of the main reasons for reprinting the volume that his experiments were a testament to allowing researchers – rather than government or committees – to control the process and funding of research (1973, 11–13).

That is clearly not the path that most countries followed. In most places, government bureaucracies fund (and through funding largely determine the direction and oversight of) a significant proportion of research through restrictions attached to funding. This standard approach is challenged by new research methods that can avoid these restrictions. Thus, in preparation for an increasing number of unregulated research projects, it is worth considering Mellanby's work as an example of how a well-intentioned, well-trained researcher conceived of his ethical obligations, and what, from our perspective, seems ethically deficient in his work? What choices might a researcher "untrammeled" by regulations make, and how similar is that situation to unregulated research today?

In what follows, I concede Mellanby's claim (shared by others today) that sometimes regulations may interfere with discoveries that a dedicated researcher, free to follow where the science leads her, could otherwise make.[7] But it is not obvious that on balance, untrammeled research leads to better results and less harm and exploitation. In fact, in some ways the lessons from Mellanby's work all share one main characteristic: that alone, as "untrammeled" researchers, each of us can be oblivious to our own biases, which can affect both the ethics and science of research. By making themselves the "measure of all things," including ethical appropriateness, researchers working without the benefit of oversight have regularly and predictably abused people involved in research. I am not claiming that these lessons represent insurmountable barriers to the ethical conduct of unregulated research, but they do need to be attended to in a systematic way in order to protect volunteers and maintain trust in the work.

Lesson 1: "Untrammeled" researchers can be ignorant of their own biases.[8]

It is unremarkable to note that each of us has biases that can affect how we approach our work and frame our hypotheses. The trick of bias, though, is that we

[6] A Canadian sit-com, "Testees," was even based on this job opportunity and aired on the FX network in 2008. For some accounts of "guinea-pigging," see Abadie 2010; Elliott 2008; and Romm 2015.

[7] However, see Sarewitz (2016) for a pointed criticism of this approach.

[8] Some new approaches will incorporate oversight; for example, in collaborative research, the entire community might have input in and oversight of the project. But there is not currently any mandate for oversight in these new methods, so my interest is in establishing the kinds of ethical issues that are missed if researchers effectively work alone, so that they can be addressed in new research.

often do not see what we do not see. Even when we accept that we are riven with biases, seek them out, and try to address them, we will still miss and fail to address some of them. Moreover, the pretense of objectivity and neutrality we may find in the "untrammeled" researcher makes it more likely that they lack the insight and humility to recognize the troubling historical record of human subject abuses and voluntarily seek out opinions that would challenge these biases.

For example, Mellanby's scabies research project included IQ tests. As he put it, "[we] put most of the patients through an intelligence test, as we were trying to find if any particular type of individual was especially liable to the disease" (84). They also included a prohibition on bathing in some aspects of the scabies research, to "rule out" the notion that scabies was a "dirt disease" (96). Mellanby's own words show that he justified these experiments by reference to "ruling out" the truth of an unexamined stereotype that perhaps could have been challenged more directly and without risk to human subjects.[9]

Just as there is a difference between a logical argument and a sound argument, there is a significant difference between research questions that are, strictly speaking, scientifically sound and research questions that are the right ones to be asking. For example, there may be a perfectly scientific way to study whether rats prefer pink or purple, but what we care about is whether and why such research is worth pursuing in the first place. Our biases can distort the scientific questions we ask, making them appear plausible only because they uncritically adopt prevailing or self-ratifying norms. Oversight of research is *meant* to "trammel" researchers. We want them to respect the norms society has crafted to govern itself, and pursue the questions to which members of that society want to know the answers.[10] A significant challenge in new, unregulated research is not just the harms that might be inflicted on the way to answering legitimate scientific questions, but also the harms that might result from asking specious questions to begin with. Regulations subject research proposals to external review to help address such issues; without the

[9] His justification also has unfortunate resemblance to one justification for the Tuskegee study, to "rule out" the notion that syphilis affected whites and blacks differently. In both cases, the only justification for the hypothesis that something needs to be "ruled out" relies on unexamined stereotypes. As Mellanby points out subsequently, being less clean had no effect on scabies infestation, but did increase susceptibility to secondary infection after scratching dirty skin (which came from not allowing volunteer subjects to bathe, not from any inherent desire on their part not to bathe).

[10] Daniel Sarewitz explored this idea to great effect in "Saving Science," concluding:

> Advancing according to its own logic, much of science has lost sight of the better world it is supposed to help create. Shielded from accountability to anything outside of itself, the "free play of free intellects" begins to seem like little more than a cover for indifference and irresponsibility. The tragic irony here is that the stunted imagination of mainstream science is a consequence of the very autonomy that scientists insist is the key to their success. Only through direct engagement with the real world can science free itself to rediscover the path toward truth (2016).

Sarewitz's concern is the frivolous questions that "untrammeled" researchers may sometimes ask, rather than potential harm to subjects, but both problems can stem from researchers untrammeled by external constraints.

regulations, researchers using new methods will need to be deliberate in evaluating how their biases may be contributing to their research.[11]

Lesson 2: Research without oversight favors a liberal interpretation of autonomy and the requirements for consent.

We want our researchers to be eager to learn and willing to test some boundaries in the pursuit of better understanding the world. Yet it can be difficult to distinguish between merely testing and completely overstepping those boundaries. The interpretation of when a volunteer is autonomous and capable of agreeing to particular interventions is a case in point. Mellanby mentions several times that he had to stop the conscientious objectors who were his subjects from going too far with experiments and endangering themselves. He does not reflect on the difference between the stopping point that *he* found prudent and those the volunteers seemed willing to tolerate, but this example raises the question of how to interpret the autonomy of these conscientious objectors during a time of conscription.[12] This is a particularly important question given Mellanby's description of his own autocratic approach to controlling some aspects of the volunteers' lives: "The volunteers were employed primarily as subjects for experimentation, and it was understood that they must always be available when necessary for this purpose" (1973, 50). There is an inherent conflict of interest when an experimenter also evaluates and interprets the autonomy of subjects, with no enforced standards of how to evaluate someone for decision-making capacity.

This liberal interpretation of autonomy can also contribute to a liberal interpretation of informed consent. Mellanby does describe having more detailed and frank conversations with his participants about the details of the experiments than most researchers would today. But what he gives with one hand he seems to take away with the other. For example, this account starts out reasonably well: "...at the outset I imagined that it would be a necessary safeguard for each man to sign some sort of detailed contract, setting forth his duties and the risks he was taking...." However, we quickly learn the true nature of his concern in what he says next: "...in order to cover me in an emergency and to have something to which to hold a recalcitrant subject if he proved non-co-operative. When I made some investigations into the legal side of the question I gathered that certain risks were being taken by me in using human volunteers" (1973, 51). Mellanby's concern here is to protect *himself* due to the risks he was taking; the "detailed contract" shifted the burden of identifying problems to his subjects and away from himself. And though he did inform volunteers about the process and possible risks of the experiments, this was motivated

[11] Obviously oversight is no guarantee that such biases may be caught, but it may make detecting them more likely.

[12] For an extended examination of volunteering for experimentation during WWII, see Newlands (2013).

in part by the goal of encouraging their compliance with the experiment: "Provided that the volunteers look upon the experiments as their experiments, and they are not treated as mindless guinea pigs, it is not difficult to get complete cooperation" (1973, 71).

These ethical issues stem from asymmetrical power between researcher and subject. It may be thought that in new forms of research like citizen or other collaborative science, this will be less of an issue due to the equal status and cooperative nature of the research, but this is incorrect for several reasons. First, much of this research is still conducted with trained scientists, even if they are not employed at an institution and/or covered by regulations regarding human subject research. Even in research deeply grounded in a community, there is often one person – sometimes even a member of that community – who has different training than other participants, and thus may have more power relative to the research project.

Second, even in research groups without structural power asymmetries (for instance, a group of co-equal biohackers dedicated to open sharing of results, or a community engaged in pursuing a collaborative research goal), it is common for individuals to try rhetorically to win arguments or earn social capital, respect, or authority. What might not be possible via structural power might therefore be possible via rhetorical talent; violent charismatic leaders have frequently gained power on this basis.

Finally, even if we postulated research teams with good intentions, it is wise to consider the possibility of the bad actor. It is not a stretch to imagine that unregulated research might be an attractive "work-around" option to those who find the current regulatory mechanisms burdensome, or to imagine that the field might be an ideal site for other kinds of co-optation. If undermining certain research approaches would be beneficial – consider citizen environmental research, climate research, or generic drug self-testing, for example (which could threaten some for-profit industries) – some parties will have reason to sabotage these approaches. Honest actors in these new research areas should consider how to protect human participants and the integrity of their work.

Like Mellanby's work, some new research would structurally favor a liberal interpretation of autonomy and "consent," whatever that might come to mean when not stipulated by regulations. To the extent that public trust and ethical requirements depend on treating people as autonomous individuals who knowingly participate in research, unregulated research involving human beings will have to grapple with ways to ensure and protect that obligation.

Lesson 3: Untrammeled researchers can be oblivious to their lack of expertise.

It is worth recalling that Mellanby was an *entomologist*. This makes sense in the context of studying the life cycle of scabies mites, but fundamentally, his research was designed to foster human combat-readiness in the context of the war. The real focus of his research was human beings: He studied how scabies was transmitted to them, what happened to the human body during an infestation, and what treatments were effective. Yet he possessed no medical training or anything that might have helped him with what were effectively clinical decisions. Perhaps he had advice and help from medical collaborators, but if so, this is not mentioned in the volume. Instead, one gets the sense that Mellanby found himself perfectly adequate to all tasks, thereby missing situations in which medical training might have been crucial – for example, when the volunteers contracted secondary infections due to their constant scratching. His framing of the project as being about scabies missed the fact that more importantly, it was about *people.*

Researchers may misunderstand the expertise required for a particular project or overestimate their own expertise in two ways. First, they might not see an ethical issue due to their role (e.g., entomologist vs. physician) or framing of the issue (e.g., a scabies problem vs. a human medical condition), and second, they might not have the expertise in the field that would help them recognize current best practices outside of their field. This is why one of the Nuremberg Code's stipulations is that researchers should be scientifically qualified to conduct the experiment. This blind spot regarding expertise is even more likely in the kinds of interdisciplinary research that citizen science and other unregulated research tend towards, because of the multiple types of expertise required and the lack of sufficient budget to recruit and incorporate others with appropriate training. Even the best-intentioned, most well-rounded researcher likely cannot identify all possible ethical issues in a project. Oversight of research will not necessarily result in identifying them all, but may makes detection more likely. In addition, the process of oversight can demonstrate ethical intention via greater transparency, thus building trust with a larger community. The "untrammeled" new researcher will have to work intentionally to find out what she does not know that she does not know.

Lesson 4: Researchers may have much to learn about truly collaborative research.

New forms of research, especially citizen science or environmental studies within communities, often depend on – or more importantly, stem from – community concerns and research goals. However, community members may not have the requisite expertise to define the research question, formulate the project, or execute the study.[13] Conducting this research may therefore require collaboration with trained researchers. However, many researchers have not been trained in appropriate methods of collaborative research, and can easily default to approaches *they* think are collaborative but others do not.

Consider this passage from Mellanby:

> One thing only was essential in the managing of the volunteers in these experiments, and that was to be willing to take the trouble to explain in detail just what was the purpose of anything which we asked them to do. I always made a practice of holding periodical meetings which they all attended, and where I gave a progress report and answered the questions which they had to ask. It is not always easy for the scientific research worker to explain to an audience which…has essentially no scientific training or special knowledge of the investigations, but I believe that it is a useful experience and leads to clearer thinking on his part as well as theirs. Also the questions asked, though often not very profound, not infrequently

[13] It is important to note, however, that many times, community members *do* have the requisite scientific expertise or can successfully learn it.

> proved useful in suggesting new lines for study. I think we usually managed to maintain a friendly atmosphere where the volunteers knew that they could come and discuss any matter concerning the research perfectly freely. As a rule, I took the volunteers entirely into my confidence about the progress of the work, though on occasion it was necessary to conceal some information, particularly when investigating such subjects as individual reactions to the presence of the causative mites and other topics with a considerable subjective element in them (54).

On one hand, this seems like quite progressive human subject research. I am aware of no researchers who conduct this kind of regular, frank discussion during a project with participants. In part this is for the reason Mellanby gives, that sometimes the empirical task requires subjects to remain ignorant of certain features of the research. However, there is significant progress that could be made in many projects towards transparency with participants without endangering the science, and Mellanby's approach is commendable in this regard.

On the other hand, one must remember that these volunteers had no recourse if they felt mistreated; Mellanby's willingness to entertain questions was a one-sided arrangement. This is made abundantly clear in Mellanby's paragraph immediately following the one above:

> Although the work was always discussed with the volunteers, there was never any question regarding the final authority in planning what was to be done. In this I was completely autocratic. Later experience has shown that this is much the most satisfactory way to run work of this kind. It is best for all concerned for one person to be in the position to make every decision that is required; of course, he should take all advice and guidance that is available, but the thing to avoid is the delegation of authority to a committee.... It is very valuable to have scientific committees to discuss scientific results and their practical application...but they are not a suitable means for the actual running of an experiment; this is the one field where a Dictator is required! (54–55)

Several other features of these passages bear mention. Notice that Mellanby seems to focus on the extra lengths these conversations require *from him*, calling the experience "taking the trouble" and noting that it was "not always easy." And, to the extent that he noted minor benefits from this practice apart from better compliance, he grudgingly notes that it can "not infrequently" lead to helpful suggestions even if they are "often not very profound." One gets the sense that Mellanby is engaging in a charade for the sake of better science, and that from his perspective, this charade only accidentally and occasionally makes the work better.

It is difficult, as Mellanby points out, to conduct scientific research by committee. Yet it can also be quite problematic for a lone wolf to run things autocratically. New, unregulated forms of research will each have to navigate these waters in ways distinct to their context. For example, patient-led research or citizen science may sometimes be science by committee, but this can produce better science and lead to results with direct applications for the communities involved. Communities engaging in this work will need to establish what collaboration means to them, rather than being forced to accept the version of collaboration that a powerful lone wolf researcher mandates.

Lesson 5: Lack of oversight favors "efficiency" approaches to research with human beings

Mellanby clearly prioritized fiscal responsibility and efficiency, a valuable characteristic in a government-supported researcher. However, too much focus on efficiency can make it tempting to try to "optimize" the use of subjects. For example, Mellanby expresses interest in efficient use of conscientious objectors by enrolling them in multiple studies, since they were already living at the site and the original experiments did not take all of their time: "I decided that we were not making the maximum use of our opportunities. Here we had a group of intelligent and willing human volunteers, living under closely controlled conditions ideal for experimental work on a variety of subjects" (1973, 68). They enrolled subjects in dietetic experiments, which he described thus: "These tests all ran parallel with the scabies work, and showed how the maximum amount could be extracted from the volunteers if they were willing to co-operate" (1975, 192). Mellanby's persistent focus on maximization of experimental subjects generates worries about how he treated them. There is nothing *prima facie* wrong with efficiency, but the natural logic of pursuing it can lead to predictable problems. A focus on efficiency can deflect our attention from questions about the need for the research in the first place, which is often the most important question. As soon as we positively value efficiency, we invite tradeoffs between it and other values – for example, the ethical treatment of subjects.

Here's how he described the addition of new research projects using his scabies subjects:

> When we first planned the scabies work we told the volunteers that they would be infected in various ways (for we were concerned mainly with transmission), but it was understood that as a rule when infection had occurred they would almost immediately be cured. I now found that I wished to follow the course of infections which continued for much longer periods. When this was explained to the volunteers they all expressed complete willingness to submit to long periods of infection, although this went far beyond the original experiment (95).

Today, we would consider this problematic for several reasons, not least because of its resemblance to the "observe the natural progression of disease" justification for the Tuskegee or Guatemalan syphilis experiments. Bearing in mind the extreme prejudice against conscientious objectors in England during both world wars, which severely curtailed their employment and social options, it was problematic that the only people to whom Mellanby put the request for additional experiments was to a group of potentially vulnerable subjects themselves. The combination of a focus on efficiency and a liberal interpretation of autonomy and consent makes it more likely that a researcher could mistreat human subjects. In unregulated research, these twin temptations could be problematic in themselves for subjects of research, and generate mistrust by potential consumers of the research.

These lessons gleaned from Mellanby's work stem from acknowledging the inherent epistemic limits of human beings. There is no shame in admitting that we need the perspective of others to see what we had previously missed – or what we have a vested interest in not seeing. Regulations are designed to force people who might not otherwise seek such critical input to submit to it. In the absence of regulations, new forms of research need to find ways of benefiting from oversight relative to their own contexts. Given what we know about human nature and history, it would be wrong to pretend that one's good intentions are enough, and that one does not need the perspective of others.

In light of these lessons from Mellanby's work, and the similar, potentially "untrammeled" nature of new forms of research, it will be important for participants in and users of the results of new research to consider the ethics of the work. The antidote to problems in untrammeled research is obviously to trammel it, by seeking opportunities for external review and assessment of the work. This does not mean merely recapitulating existing regulatory structures. In fact, this is one of the most promising, interesting, and exciting aspects of new research approaches, because – untrammeled by current regulatory principles and requirements – new researchers can make different, and potentially better, choices about what ethical goals to aim for in their work. They can also incorporate a kind of reverse oversight system, in which the subjects of the research are offered the ability to evaluate and exclude certain research or researchers from collaborative opportunities.[14] New research ethics could help to improve an ossified regulatory system of ethics built for a different era.

Conclusion

To offer such a frank account of his work, Mellanby clearly thought of it as at least honorable, if not downright exemplary. *Human Guinea Pigs* is written as though there is nothing about the experiments to hide, reflect critically on, or publicly regret. This in itself is revealing: not only did Mellanby think there was nothing exceptionable about what he did, he seemed fairly confident that no one else would, either. It is particularly remarkable that he made no substantive changes to his account in the second edition, despite having been present at the Nazi Doctor's trial, and despite the fact that the Nuremberg Code was articulated over 25 years prior to the appearance of the second edition.[15]

[14] For example, see Schneider (2015) for "Turkopticon," a browser add-on that allows individuals doing work for Amazon's Mechanical Turk site to review and exchange information about individuals requesting piecework on the platform. Also see Grayson et al. (2019) for an account of how Sage Bionetworks employs a "reverse panopticon" to monitor researchers' use of data.

[15] Interestingly, however, in a paper published shortly after the second edition of *Human Guinea Pigs* appeared (based on a lecture given to the British Association for the Advancement of Science), Mellanby focused specifically on voluntariness in his scabies work: "From the outset I was adamant that I would only use genuine volunteers.... I was not prepared to accept 'volunteers' who were directed to work as guinea-pigs, but only individuals given unconditional exemption or a wide range of humanitarian choices. The proposed type of work was always full explained, and they were assured that they could withdraw at any time" (Mellanby 1975). This maps very clearly to the requirements of the Nuremberg Code regarding voluntariness, full information, and ability

Even from our present regulatory and ethical standpoint, Mellanby's work *does* have aspects worth celebrating. There were collaborative features of his work that many researchers today would feel unnecessary, but that Mellanby seems to have felt were required. But even here, his reliance on his own perceptions of the work, seemingly without seeking critique or objection from others, demonstrates the potential dangers of new methods of unregulated research with human subjects. Consider the following:

> The volunteers had been engaged for work on transmission and had been told at the outset that their infections would be treated immediately on diagnosis. However, our work had revealed so many gaps in knowledge that we wished to prolong infections and study the process over longer periods. This was explained to the volunteers, and every single one expressed his willingness to submit to prolonged infestation with the itch mites. In some cases this lasted for months; men got little sleep and lacerated their itching skins with their finger nails, but never did a volunteer ask to be cured before the experiment was concluded. I think they were encouraged because they could see the significance of their work, which resulted in better diagnosis, treatment, and protection for millions of other people who might otherwise have suffered. They were always kept fully informed and were allowed to feel that they were the colleagues of the scientists in charge, as well as their experimental subjects (1975, 191–192).

Mellanby gives only his own assessment of what the volunteers were *probably* thinking, exhibiting no interest in discovering or recounting what they *actually* thought, even when they had suffered from itching and lost sleep for months, and when the experiments exploited the pressure the conscientious objectors were under to contribute to the war effort. The risks and harms were real: According to one of Mellanby's colleagues, "a volunteer receiving a depleted intake of vitamin C…died of a heart attack" (Weindling 1996), and another volunteer was hospitalized during the Vitamin A experiment: "He didn't feel ill, but noticed that his night sight had failed. After 18 months [on the diet], however, 'I suppose nature caught up with me. I was taken to hospital. When I came out I felt like an old man, not someone of 25. It was the end of experiments for me,' he said with regret" (Peace Pledge Union n.d.). One of the volunteers later articulated the "voluntary" circumstances under which he came to join the scabies experiments: "When the war broke out [Norman Proctor] was a baker, a reserved occupation. 'The other men refused to work with me because I was a concscientious objector so I got the sack. I was out on a limb. Then I heard about Dr. Mellanby's experiment on scabies. I thought it would help others'" (Collins 2006). Like Hickens, Proctor lost his night vision during the Vitamin A deficiency experiment. He recovered once he left the experimental diet,

---

to withdraw. This description of his motives may well have been true, but it was not described this way in the first edition. The word "withdraw" does not appear at all in either edition, and where the word "information" appears, it is almost always about information gleaned, not information disclosed. One use of the word involves concealing information from participants. At the very least, this demonstrates that disclosing to readers of scientific work that participants were treated as autonomous agents was not important at the time of either edition. The timing of this article hints that perhaps Mellanby felt it necessary, after publication of the second edition, to elaborate on the voluntariness of the experiments.

but noted that "two volunteers got TB, one almost certainly due to the experiment," while others suffered bleeding gums and scurvy.

In 2006, John Pemberton, one of the surviving research workers, sought out surviving participants to canvass their opinion, over a half-century later, on their participation in Mellanby's experiments. Beginning from a list of 23 volunteers recalled by Pemberton and Proctor, they found that "17 were known to have died, three were untraced, and three replied to my short questionnaire. Those who replied all thought that the experiments had been very worthwhile and stated that they would have volunteered again if required" (a very interesting sentence construction) (Pemberton 2006, 556). Of course, this small sample, taken so many years later, does not give us confidence that all volunteers were of the same opinion. However, what is clear is that researchers using new methods not subject to existing regulations are morally obligated to give serious thought to what constitutes voluntariness, autonomy, capacity to consent, risk-benefit ratios, and a host of other familiar values in human subject research. One of the most significant dangers in new research is that well-intentioned researchers will assume that their own good intentions suffice, when history abundantly demonstrates that they will not.

References

Abadie, Roberto. 2010. *The professional guinea pig: Big pharma and the risky world of human subjects*. Durham: Duke University Press.

Beecher, Henry. 1966. Ethics and clinical research. *New England Journal of Medicine* 274 (24): 1354–1360.

Collins, Alice. 2006. Commentary: Guinea-pigs' private war. *International Journal of Epidemiology* 35: 558–560. [Originally published in the Sheffield Telegraph, August 29, 2003, p. 12.] Available online: https://academic.oup.com/ije/article/35/3/558/735688.

Cooper, Melvin Wayne [Editor and translator]. 2019. *Medical Jurisprudence and rules of the medical profession*, by Rudolph Ramm. Translation with Introduction. Dordrecht: Springer.

Currier, R.W., Shelley F. Wlton, and Bart J. Currie. 2012. Scabies in animals and humans: History, evolutionary perspectives, and modern clinical management. *Annals of the New York Academy of Sciences* 1230: E50–E60.

Editor. 1947. "Dr. K. Mellanby, O.B.E." *Nature* (May 24), 702.

Elliott, Carl. 2008. "Guinea-Pigging," *The New Yorker* (December 30). Available online: https://www.newyorker.com/magazine/2008/01/07/guinea-pigging.

Engber, Daniel. 2012. "Test-tube piggies: How did the guinea pig become a symbol of science?" *Slate* (June 18). Available online: https://slate.com/technology/2012/06/human-guinea-pigs-and-the-history-of-the-iconic-labanimal.html.

Farmer, Ben. 2014. "British soldiers' WW1 trench battles with STDs, rheumatism and wasp stings," *The Telegraph* (October 8). Available online: https://www.telegraph.co.uk/history/world-war-one/11148580/British-soldiers-WW1-trench-battles-with-STDs-rheumatism-and-wasp-stings.html.

Grayson, Shira, Christine Suver, John Wilbanks, and Megan Doerr. 2019. *Open data sharing in the 21st century: Sage Bionetworks' qualified research program and its application in mHealth data release* (December 11, 2019). Available at SSRN: https://ssrn.com/abstract=3502410 or http://dx.doi.org/10.2139/ssrn.3502410.

Jackson. 2004. Scabies by Kenneth Mellanby. *Journal of Cutaneous Medicine and Surgery* 8 (2): 73–76.

Maibach, Howard J. 1974. "Mellanby's Scabies," (Letter to the Editor). *Archives of Dermatology* 109 (3): 409. Available online: https://jamanetwork.com/journals/jamadermatology/article-abstract/533867.

McCullough, Laurence. 1998. *John Gregory and the invention of professional medical ethics and the profession of medicine*. Dordrecht: Springer.

Melicharova, Margaret. 2006. "The interest of science. Scabies and conscience," *Peace Matters* 52(Autumn). Available online: http://archive.ppu.org.uk/peacematters/peacematters/2006/2006_autumn1.html.

Mellanby, Kenneth. 1973. *Human guinea pigs*. London: Merlin Press.

———. 1975. Experiments on human volunteers. *Journal of Biological Sciences* 7: 189–195.

———. 1988. "Itching to study lice and mites," *The Scientist* (May 2). Available online: https://www.the-scientist.com/news/itching-to-study-lice-and-mites-63001.

Mellanby, John. n.d.. "*Mellanby history*." Available online: https://sites.google.com/site/mellanbyhistory/mellanby-family/scientists

Newlands, Emma. 2013. 'They even gave us oranges on one occasion': Human experimentation in the British Army during the Second World War. *War and Society* 32 (1): 19–63.

Peace Pledge Union. n.d.. "*Bernard Hicken's story*." Available eonline: http://archive.ppu.org.uk/learn/infodocs/cos/st_co_wwtwo6.html.

Pemberton, John. 2006. Medical experiments carried out in Sheffield on conscientious objectors to military service during the 1939–1945 war. *International Journal of Epidemiology* 35: 556–558. Available online: https://academic.oup.com/ije/article/35/3/556/735661.

Perring, Franklyn. 1994. "Obituary: Kenneth Mellanby," *The Independent* (January 11). Available online: https://www.independent.co.uk/news/people/obituary-kenneth-mellanby-1406138.html.

Romm, Cari. 2015. "The life of a professional guinea pig," *The Atlantic* (September 23). Available online: https://www.theatlantic.com/science/archive/2015/09/life-of-a-professional-guinea-pig/406018/.

Sarewitz, Daniel. 2016. Saving science. *The New Atlantis* 49: 4–40. Available online: https://www.thenewatlantis.com/publications/saving-science.

Savin, J.A. 2002. Mellanby on scabies. *Clinical and Experimental Dermatology* 27: 86–87.

Schneider, Nathan. 2015. "Intellectual piecework," *Chronicle of Higher Education* February 16.

Weindling, Paul. 1996. Human guinea pigs and the ethics of experimentation: The BMJ's Correspondent at the Nuremberg Medical Trial. *British Medical Journal* 313 (December 7): 1467–1470.

Index

L. M. Rasmussen (ed.), *Human Guinea Pigs, by Kenneth Mellanby: A Reprint with Commentaries*, Philosophy and Medicine 134,
https://doi.org/10.1007/978-3-030-37697-0

FSC
www.fsc.org
MIX
Papier | Fördert gute Waldnutzung
FSC® C083411

Kolibri 360 GmbH
Ferdinand-Jühlke-Straße 7
99095 Erfurt, Deutschland
produktsicherheit@kolibri360.de